TAKE ME *to the* RIVER

To Sharon,
Come canoe the river!
Coke

Fishing, Swimming, and Dreaming on the San Joaquin

Edited by Joell Hallowell and Coke Hallowell

Assistant Editor: Donna Mott

A project of the San Joaquin River Parkway and Conservation Trust, Inc.

Heyday, Berkeley, California

Heyday would like to thank the James Irvine Foundation for their support of Central Valley literature.

Library of Congress Cataloging-in-Publication Data

Take me to the river : fishing, swimming, and dreaming on the San Joaquin / edited by Joell Hallowell and Coke Hallowell ; assistant editor, Donna Mott.
p. cm.
"A project of the San Joaquin River Parkway and Conservation Trust, Inc."
Includes bibliographical references.
ISBN 978-1-59714-143-7 (pbk. : alk. paper)
1. Naturalists--United States--Biography. 2. San Joaquin River Watershed Region (Calif.)--Description and travel. 3. Parkways--California--San Joaquin River Region. I. Hallowell, Joell. II. Hallowell, Coke. III. Mott, Donna.

QH26.T35 2010
508.794'8--dc22 2010007152

Cover Art: A summer outing at the beach at Skaggs Bridge. Brooke Wissler's aunts Ethyl Fleda and Louise Mordecai with a family friend, circa 1910–1920. Courtesy of Brooke Wissler.
Book Design by Lorraine Rath
Printed in Korea through Tara TPS through Four Colour Print Group

Orders, inquiries, and correspondence should be addressed to:
Heyday
P. O. Box 9145, Berkeley, CA 94709
(510) 549-3564, Fax (510) 549-1889
www.heydaybooks.com

Right: At Fresno Beach, at Lane's Bridge, circa 1919. Courtesy of Clayton Vander Dussen

Photo on previous pages: Brooke Wissler's father and mother with friends, circa 1910–1920. Courtesy of Brooke Wissler

10 9 8 7 6 5 4 3 2 1

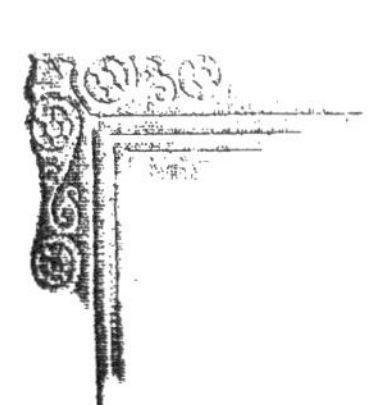

Contents

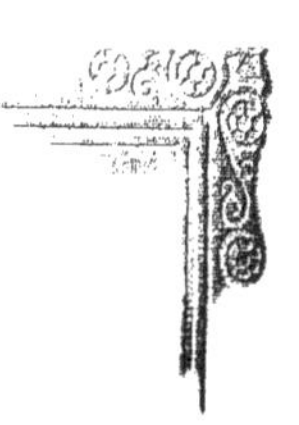

Stories

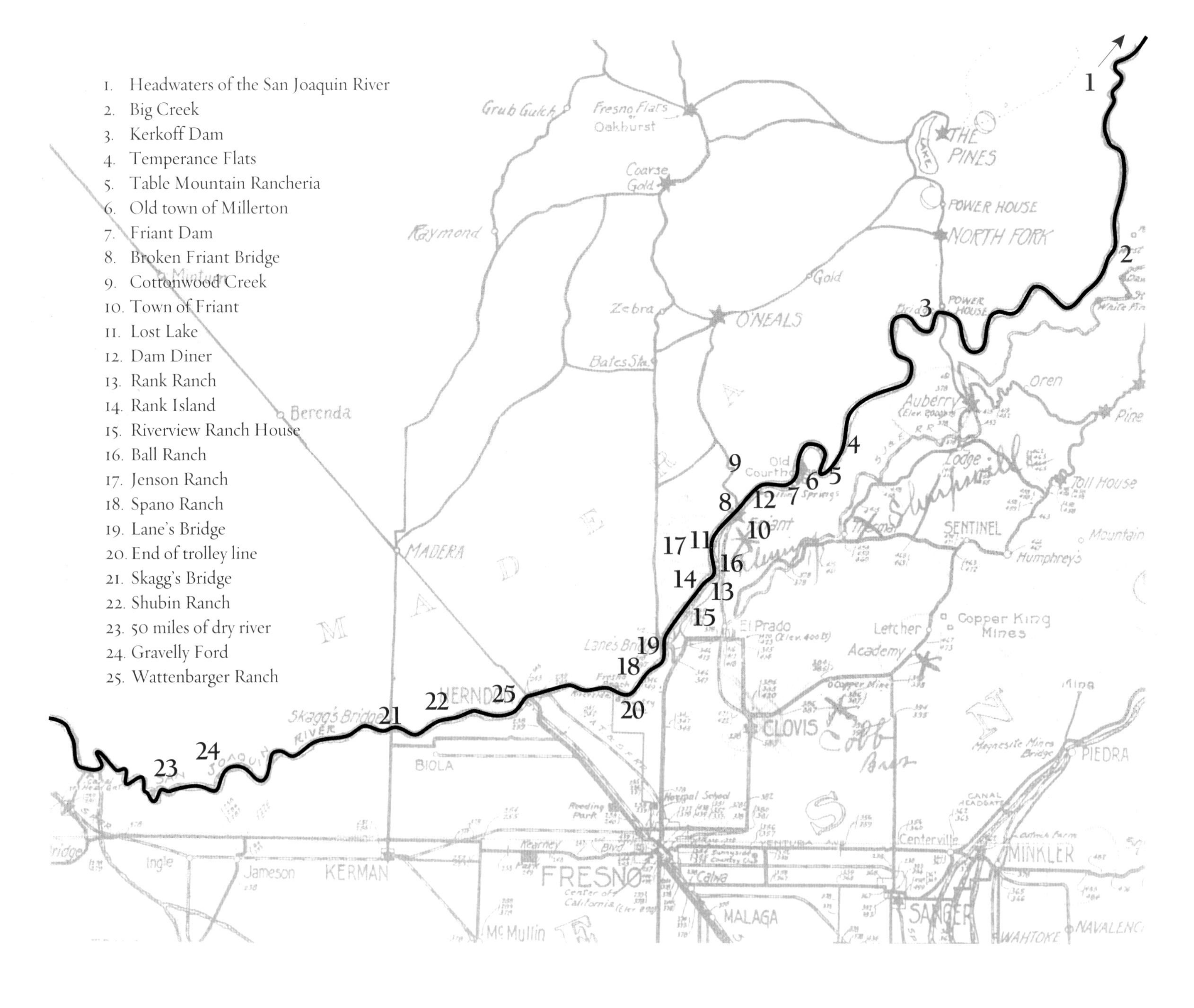
1. Headwaters of the San Joaquin River
2. Big Creek
3. Kerkoff Dam
4. Temperance Flats
5. Table Mountain Rancheria
6. Old town of Millerton
7. Friant Dam
8. Broken Friant Bridge
9. Cottonwood Creek
10. Town of Friant
11. Lost Lake
12. Dam Diner
13. Rank Ranch
14. Rank Island
15. Riverview Ranch House
16. Ball Ranch
17. Jenson Ranch
18. Spano Ranch
19. Lane's Bridge
20. End of trolley line
21. Skagg's Bridge
22. Shubin Ranch
23. 50 miles of dry river
24. Gravelly Ford
25. Wattenbarger Ranch
Grub Gulch
Fresno Flats
Oakhurst
Coarse Gold
THE PINES
POWER HOUSE
NORTH FORK
Raymond
Gold
Zebra
O'NEALS
POWER HOUSE
Bates Sta.
Oren
Auberry
Pine
Berenda
Lodge
Toll House
SENTINEL
Mountain
Humphrey's
MADERA
El Prado
Letcher
Copper King Mines
Academy
Skagg's Bridge
RIVER
SAN JOAQUIN
CLOVIS
BIOLA
PIEDRA
Roeding Park
Normal School
Centerville
Kearney Blvd
MINKLER
Ingle
Jameson
KERMAN
FRESNO
Center of California
MALAGA
SANGER
McMullin
WAHTOKE
NAVALENC

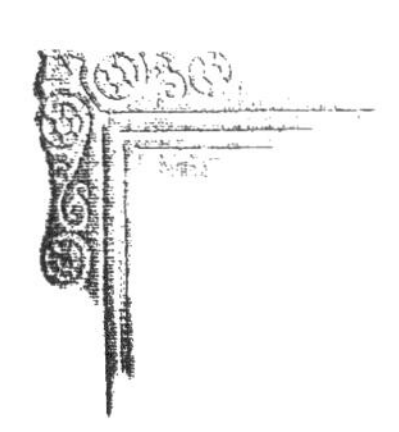

Foreword

A Primer to the San Joaquin River

Dave Koehler

Executive Director, San Joaquin River Parkway and Conservation Trust

EVERYONE LOVES A GOOD STORY, although we hope this collection will take you beyond the point of just pure enjoyment and into a state of discovery. For me, stories of the river are fundamental to my passion and work to preserve the San Joaquin River. My boyhood years in Fresno during the 1960s were devoted to fishing at the river. With my bike leaning on an old bridge pylon, I daydreamed and gladly fished the mornings away. When I wasn't fishing, I was reading up on fishing techniques and getting my tackle ready for the next trip. I listened in earnest to the stories my grandfather told me about the salmon runs, and to the stories my dad told about picnics with my mom and Sunday drives to the river in his 1932 roadster. During college, it was the stories told by field biology professors that further developed my interest in conservation.

The stories in this book are a window into the relationship between land and people. The twentieth century was the era of the San Joaquin Valley's most explosive growth in population and infrastructure. The threads stitching these stories together are the bed and banks of the San Joaquin River. Each storyteller provides a unique, focused lens through which to view life on the river's shores, and the voices call us to discover our own relationships to the natural world.

I believe the depth of these stories can best be appreciated with some knowledge of the river's natural and cultural history. The San Joaquin River's headwaters originate high in the Sierra Nevada, spanning the area between Yosemite and Kings Canyon National Parks. From there it travels 330 miles before it reaches the Sacramento–San Joaquin Delta and the San Francisco Bay, the largest estuary on the west coast of North America. The San Joaquin River is a vital source of clean drinking water for millions of Californians, supports thousands of acres of rich cropland, and is a life-sustaining resource for fish and wildlife, including many threatened and endangered species.

The San Joaquin once supported a vast, diverse, and healthy ecosystem. Spring snowmelts sent thunderous flows down the Sierra annually that laid down fresh layers of cobblestones, sand, and soil. The meandering river split into multiple channels and spread its floodwaters over a wide plain. Flowing water, rich soil, and the bulwark of plant succession gave rise to an array of wetlands and streamside forests. Journals of early explorers document that wildlife in the San Joaquin Valley in 1850 included herds of tule elk, pronghorn antelope, black and grizzly bears, and a spring Chinook salmon run that numbered in the hundreds of thousands.

Beginning with California's gold rush, things began to change. By the 1920s, the combination of placer mining in the river's tributaries, water diversions, and draining wetlands in the lower reaches had altered the river's navigability enough to put an end to commercial steamboat travel between Stockton and Fresno. Also in that timeframe, harnessing the river's energy through hydropower became a high priority in order to support California's growth. A series of dams, tunnels, and penstocks was appended to the river's course as it carved its way down the granite batholiths of the Sierra.

Perhaps the greatest single change to the river's natural history was the building of Friant Dam, under construction from 1937 to 1947. The dam is located on the river's main stem at the base of the Sierra foothills and the historic town of Millerton. Friant Dam is 319 feet high, 3,488 feet long, 293 feet thick at the base, and contains 2,135,000 cubic yards of concrete. The dam is one of twenty dams that are part of the Central Valley Project, which was authorized by Congress in 1933 and is managed by the U.S. Bureau of Reclamation. The purpose of the Central Valley Project was primarily to capture and store water for distribution to California's developing agricultural industry, and secondarily to provide flood protection. Most of the San Joaquin River's water is diverted at Friant Dam into two canals. The majority of its water now travels down the east side of the valley through the Friant-Kern Canal, which is 152 miles long and supplies water as far south as the Kern River near Bakersfield. The smaller Madera Canal is 36 miles long and supplies water to Madera County.

Building Friant Dam resulted in dewatering sections of the San Joaquin River and caused a portion of its bed to go dry. The river's dewatering significantly contributed to the collapse of the Bay-Delta ecosystem. It put an end to an important salmon run and has impacted California's commercial fishery. The resulting loss of wetlands and streamside forests has endangered certain species of plants and animals. In the lower reach of the river, between Los Banos and Merced, sections have been entirely lost and the river's floodwaters, when they come, are shunted into a series of canals and bypass channels.

It's been said that the San Joaquin River is one of the most altered and litigated rivers in California. The outcome of two historic lawsuits significantly affected the entire river. The first was a lawsuit filed in the 1940s as concrete for the dam was still being poured. The suit was brought by Everett G. Rank, Sr., representing his family and others living downstream of the dam, against U.S. Secretary of the Interior Julius Krug. Their complaint focused on the Bureau of Reclamation's intent to shut all the water off below the dam, water that the farming families relied upon for their crops. After sixteen years of court wrangling, *Rank v. Krug* went all the way to the U.S. Supreme Court, where in the end very little was actually decided. The upshot of all the court and administrative proceedings resulted in the bureau's providing just five cubic feet per second of flow at Gravelly Ford, which is a naturally porous spot on the river about thirty-eight miles downstream of Friant Dam. The farmers didn't get all the water they asked for, nor was it enough to maintain a living river and its salmon run.

The Natural Resources Defense Council (NRDC) and thirteen other conservation and fishing organizations filed the second of the historic lawsuits in 1988. It was brought against Kirk Rogers, regional director of the Bureau of Reclamation. As the bureau's forty-year contracts for Friant water delivery came up for renewal, NRDC's complaint focused on the unmitigated damage to the river's fishery that the building of the dam had caused. Again, years of court wrangling ensued and the case of *NRDC v. Rogers* was ultimately presented to the U.S. Supreme Court, which remanded it to U.S. District Court Judge

Lawrence K. Karlton in Sacramento. It was ruled that the bureau had illegally dried up the San Joaquin River. However, instead of the Court deciding the specific remediation, the NRDC and Friant Water Authority jointly crafted a river restoration plan to settle the matter, which the Court approved. In March 2009 Congress passed the Omnibus Lands Act, which included the restoration settlement, and President Obama signed it into law. The law gives federal authorization for implementing the San Joaquin River Restoration Program. The program, now in progress, has two primary goals: 1) restore water flows adequate to recreate a living river with naturally producing and self-sustaining populations of salmon and other fish between Friant Dam and the Merced River confluence, and 2) develop a water management plan to help provide water supplies to support a vibrant agricultural economy. The target date for reestablishing salmon is 2014.

Alongside changes to the river's ecosystem, layers of changes in the river's cultural history have been laid down. In the 1800s, Native American people along the river suffered devastation from diseases brought by settlers and from conflicts such as the Mariposa Indian Wars. By 1900 only a small number of people from the Dumna tribe remained along the river in the valley, and only a slightly larger number of people of the Mono tribe lived along the river's banks in the Sierra foothills and mountains. As the twentieth century began, waves of various groups of European immigrants were arriving, settling in the San Joaquin Valley and focused on developing the valley's agricultural industry. By 1920 Fresno was in its glory days as a hub of agricultural commerce. In the 1980s, a new wave of immigrants from Southeast Asia began settling in the valley, families from Vietnam and Laos that fled their homes in the aftermath of the Vietnam War. The Hmong people in particular came to the valley to farm. Throughout California's history, people originating in Mexico and other Latin American countries have played a prominent role and have been vital to sustaining the valley's agriculture and other industries.

In 1988 the San Joaquin River Parkway and Conservation Trust was formed to create and protect the San Joaquin River Parkway—a greenway of protected open space and trails along the first twenty-two miles of the river as it flows through the communities of Fresno and Madera, below Friant Dam. Since 1990 it's been my privilege to work with so many people in the community that are dedicated to this important effort. The Trust provides opportunities for the public to connect to the river through a summer River Camp, school field trips, canoe tours, workshops, restoration workdays, nature hikes, and trail outings. Through these connections, new stories unfold.

There's so much at stake in the decade to come; river restoration, improving our water supply, addressing the impacts of climate change, and planning for the valley's growing population are just a few of the important issues that face us. Drawing upon the events of our past, it's appropriate that we ask ourselves, "What have we learned about our relationship to land as part of our community?" The voices of the storytellers in this collection spark memories of our own stories and provide us with a tool for charting a brighter future.

Viola Adlesh (right) with her uncle Jack and his wife, Amelia. Courtesy of Sally Adlesh

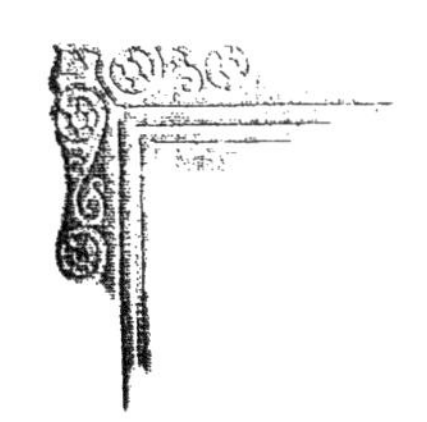

Introduction

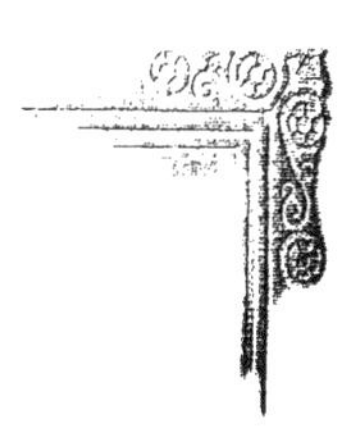

The Story of the Stories

Coke Hallowell

Chairman of the Board, San Joaquin River Parkway and Conservation Trust

WHEN MY DAUGHTER, JOELL, AND I began interviewing old-timers ten years ago, we had no grand plan. We knew that we could not set out to be official "oral historians"—neither of us had that background—so we decided to keep it simple. We had confidence that people would know how to tell their own stories in their own way. Joell brought her video camera along and I came with one question: What was life like along the San Joaquin River? We began with Dave Phillips, whose grandparents had lived on a large farm by the river. In 1894 they moved into the house that the Parkway Trust had recently acquired as our Center for River Studies. We learned that day about a fascinating piece of river history and it excited our curiosity about others who might be willing to relate their river memories to us.

We have now collected over sixty tales of life along the river. We have interviewed people with a great variety of perspectives, and gradually the stories have begun to reveal a wide-angled view of what makes the San Joaquin River what it is today. We quickly abandoned the idea of only interviewing old-timers and began to include younger generations who had lived by the river, along with water experts, environmentalists, and community activists.

Finding these storytellers was an adventure in itself. I began with a few people I knew. Everard Jones was a coach at Clovis High School who I knew was an avid fisherman. Fred Biglione was the contractor who built our home in the foothills and I was aware of his long history on the river. I asked a few neighbors to participate and a few strangers came to me with their stories. As the project escalated, one storyteller led to another. Unfortunately, we were not able to represent all of our interviews in this book, but the DVD collection is available to view at the Center for River Studies. We encourage readers to visit.

We are grateful to all of the storytellers for their time and candor. These stories capture a snapshot of river history that propels us into the future—a future that holds the promise of river restoration. I truly believe the San Joaquin River will soon become whole and healthy again. That revival will be a great gift to the people of California.

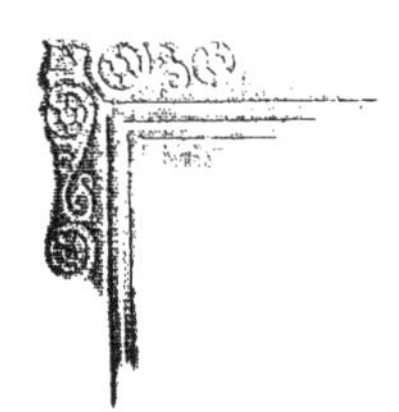
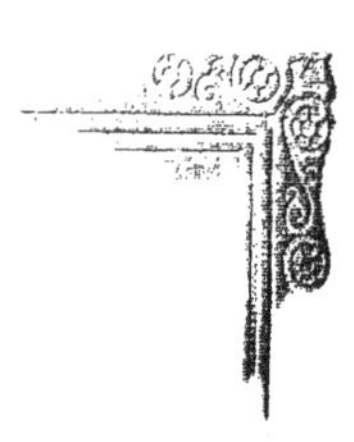

Editor's Notes

Memories and Evidence

Joell Hallowell

Story Editor

WHAT MAKES A STORY A STORY? Usually there is a plot and a theme; there are characters, conflict, and resolution—and there is a setting. Sometimes a story's location simply drifts quietly in the background; in this book, however, the landscape *is* the story. These thirty-three narratives all have one thing in common—the San Joaquin River. Some of the tales in this collection are separated by almost one hundred years, but still the river remains a constant.

Nothing is truly stable though, even the shape of the land itself. In the span of the past century, the foothills of the Sierra Nevada have been worn by winds and weather, the peaks of mountains have been flattened by bulldozers, and the San Joaquin River has continued to carve through this valley, changing it every day. Our memories are in constant flux too, and so then are the stories we tell. On a good day we remember exactly how the river smelled to us that first time we tossed a line into the riffles. Other days are fuzzier; we see our past like faded photos, grainy and indistinct.

The storytellers in this collection told us the truths of their memories. Some were in their nineties by the time we asked them to tell their childhood stories. Seventy-five years ago, when a young girl caught sight of a raucous baptism on the banks of the river, she didn't know if the participants were Pentecostal or Baptist; she only knew that it was a remarkable sight. The exact year of the big pre-dam flood is now unimportant; it was a long time ago and there hasn't been anything like it since. Throughout these tales, readers will find discrepancies and contradictions as well as collisions and corroborations. That is the nature of reminiscence, the nature of storytelling.

Photographs, however, are a miraculous safeguard against forgetting. The family photos in this collection help to freeze in time a very particular day on the beach at Lane's Bridge, one incredibly large salmon on the end of a fishing line. Those muddy boots; that big black Model T; that turnip from the garden. Those of us lucky enough to have a box of old photos in the attic not only have vivid images to help spark our memories, a place to begin, we also have evidence to back up our family stories—proof.

Often, after we finished taping the interviews for this project, we were presented with verification—an old family photograph, sometimes an album full. I was invariably amazed by the storyteller's precise and accurate recall. The barn was huge, just as described. The fish really were that big, the San Joaquin River was actually that wide. A grand river that lives on.

STORIES

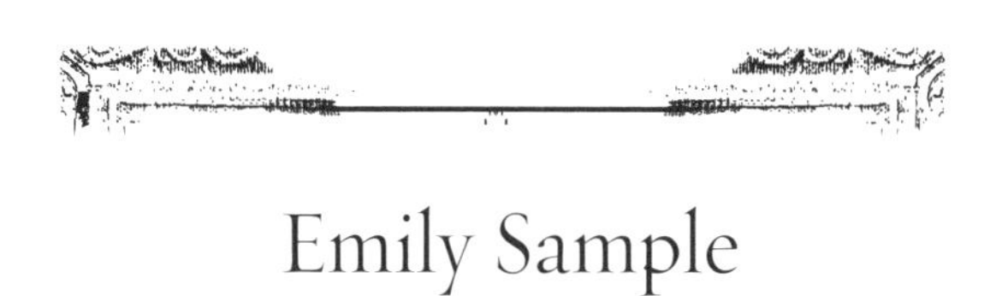

Emily Sample

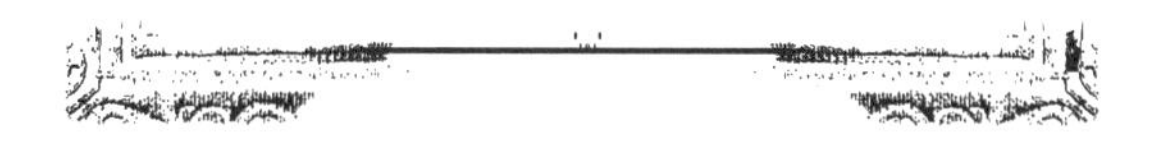

Emily Sample grew up on tribal land at Table Mountain Rancheria, just a few miles from the San Joaquin River. Her summers were spent camping along the banks of the river with other members of the rancheria community. From her home in Clovis, Emily recollected those years with affection.

I SPENT ALL MY CHILDHOOD SUMMERS by the river but I don't know if I can remember much; it was a long time ago. At Friant and at Table Mountain, it would be so hot in the summer that all the Indians would go down to the river to camp, right above where the dam is now. There were lots of willows and oaks and it was cooler there. We'd start camping by the river in June, through all of July and August, until the grape harvest started. We'd go back home to Table Mountain in September. During that time my dad and the other men would all go off to work from the campsite while the women and children stayed by the river. It was great for the kids; we really had fun. I remember seeing the salmon swim up the river; they were bright pink. And I learned to swim in the San Joaquin too. It was a big river but I wasn't afraid. I guess I didn't know the difference. I didn't know it was dangerous. I had one sister and four brothers younger than me; Bessie, Jerry, Lee, Clarence, and Paul. They've all passed away by now. I'm the only one still alive.

Indians are very resourceful; they find good ways to do things. Our mothers would cook over a campfire the whole summer and do the laundry in the river. They'd find a big, flat rock; we'd lay out the clothes, soap them up, and rub them on the rock; then we'd hang the clothes on the bushes to dry.

I guess that's why I don't like to camp now. My kids were always saying how they'd love to go camping. Well, they can have it! It didn't really seem that hard at

Emily Sample, age thirteen. Courtesy of Shirley Ramírez

The Geo Jones family camp, next to the San Joaquin River in Friant. Emily Sample's mother is seen from the back, preparing beans for cooking. Courtesy of Shirley Ramirez

the time; it was just a way of life. It wasn't a terrible hardship but I'd had enough of that by the time I moved away.

The men would come home from work and get up a party to fish for salmon at night, with lanterns, from a boat. They would go spearing all night. They could see the shine of the fish by the lantern light. At that time, it was legal for Indians to spear. But nobody else was allowed to use this method to fish for salmon.

In the morning, the women would give the men their breakfast and then work on the fish all day. The women would split and wash the salmon. They'd fillet them, cut them into long strips, lay the fillets over ropes or wires, and let the fish hang out and dry in the sun. They'd use a lot of salt so the flies wouldn't get on the fish. It didn't take too long to dry—just three or four days. When it was all done, they'd sack them up and we'd have salmon for the whole winter.

The town of Millerton was already abandoned when I was a child, except for the Charlie Green family that lived at old Fort Miller. They raised cattle there. Some Indians would work for him herding his cattle, but not my dad. My father worked on another farm down by the river. It was called the Lee Gros Ranch. It was about four or five miles below Friant. There were vineyards to work, and they raised watermelons and vegetables in the summer. They used the water from the river to irrigate the crops. There was a big old pump and during the summer they would have that thing running day and night, taking water from the river. They didn't have a tractor on the farm, so my dad would walk all day long behind the mules with a plow. It was a wonderful day when the Green family finally acquired a tractor; my dad didn't have to follow those mules anymore. I went out to the Gros Ranch years later and saw the old farming tools which had been found out in the fields. I said, "Oh my God, those are the plows that my dad used to work." There was also the old truck that he'd learned to drive; it was old and a wreck, and that first tractor was there, too—all the things my dad worked with when I was a little girl. The next time I went out to the ranch, with my daughter, they'd cleaned the place up and dumped the tools somewhere. I wish I could have gathered up some of that old farming equipment for myself.

After my father quit working on the Gros Ranch, he'd go down to Clovis and camp on somebody's farm, and from there they'd go pick grapes. At that time, all the men did that. All the Indians came down from the mountains, from North Fork and Auberry; they'd go down to work in the fields in the summer—cutting and picking peaches, and picking grapes.

During the Depression, in the 1930s, my dad would pan for gold from the river. He'd get enough gold to buy food—food wasn't very expensive in those days. He'd take the gold to the post office in Friant and they'd give him cash. Everybody panned for gold then along the river, not just the Indians but everybody who lived up there. My dad would pan with my brother, Lee, along the banks of the river on the slopes. They'd cut tunnels way back, bring out the dirt and then wash it in the river. Maybe it would be just three dollars for a whole day of panning, but that was enough for food.

I remember two big Indian events along the river when I was growing up. One was way up the river from Friant, at the Hudsons' place—they had one there, but I think that one was a wake. We had a big dinner and the white men were dancing, too, at that one. The other big event, when I was ten or twelve years old, was what they called a big fandango. They had lots of games, hand

"It would be so hot in the summer that all the Indians would go down to the river to camp."—Emily Jones Sample. Courtesy of Gene Rose

games and races, both horse racing and people racing. It was a good time and there was lots to eat, including traditional foods. That's the only fandango I remember. It lasted about two or three days and people camped during the event. To tell people about the fandango, a *winatun*, a special messenger, went out and delivered the invitations in person. Invitations went to different tribes, to North Fork or Lemoore, wherever the tribes were. I don't know how many people were there because I was just a small child, but to me it seemed like hundreds. They danced and sang Indian songs. Each tribe had their own songs but they all knew each other's songs too. I didn't learn any of the songs, but my mother used to do a lot of traditional singing. She sang at wakes all night long but she couldn't sing those songs for me. I asked her, but she said, "No, I can't do that, it's not allowed." She would never sing those songs except at one of those occasions, for they were only for particular occasions. She could sing the hand-game songs

"My mother made baskets....She got all of the things she needed for the baskets by the banks of the river." —*Emily Jones Sample*

anytime; that wasn't taboo. The hand games are played with sticks, ten of them, that are passed behind the back. There were two teams and the opposite team would guess where the stick was. If they guessed right, then that team would get the stick. All the time the players sang hand-game songs. They were very pretty songs. There were prizes for the winners—a blanket, a quilt, a sack of flour. They took the long stick called a clapper, split it, and made a rhythm with it. There are some Indian people who still play that game. They look forward to it and it's still a lot of fun.

My mother made baskets. Her name was Laura Domingo Jones and she was well known for making very beautiful baskets. She got all of the things she needed for the baskets by the banks of the river. The women used to come down below Lane's Bridge to gather white roots for the baskets along the river where it was sandy. They took roots from the sedge too. I tried to make baskets. I know how it's done but I never could do it. It's hard now to gather the materials. They used to use redbud to make the red part of the design. They gathered little tiny twigs of redbud, but it's now protected so you can't pick it anymore. Up at Table Mountain, though, they're planting some of the roots and other plants that will make it possible to keep making baskets. They use the sourberry for the picking baskets and the deer grass for tying the roots. That stuff can be very vicious to clean. When you're stripping it, you can get sores on your fingers. My mother got a bad sore once from that deer grass. She had a hard time healing it. It was hard work, but my mother really enjoyed making baskets. I remember she'd sit on the Lee Gros Ranch under one of those big umbrella trees. She sat on the ground making her baskets. It could take three or four months to make one basket, depending on how big the basket was. She worked at it every day.

I used to help my grandmother, Rosa Domingo, pound the acorn. We'd walk out there carrying all our stuff, the baskets to shake it and the rocks to pound it. My grandfather, John Domingo, would help us get all the stuff up there; then I'd stay there with my grandmother all day. I'd bring her the water and she'd fix me a pile so I could pound too. She'd have all the flour ready, then the next day we'd have a get-together to cook it. My mother and my aunts would come and whoever else wanted to help. I loved acorn; I still love it. But I never make acorn now because it's so labor-intensive. There's a lady who lives in North Fork who makes it for sale. Her name is Grace Tex. We're going to have some at my birthday party, coming up. There are two kinds and each is cooked a little bit differently. There's the soupy kind that you put in a cup and drink. The more solid kind you can eat with a spoon or even with your fingers. I like the hard kind better. I'd like to go back up and find the bedrock where my grandmother and I used to pound. I know where it is and I'm sure it must still be there—unless they blasted it away for construction.

When the dam was built, the salmon couldn't get upstream anymore. That stopped it altogether. I think the last spawning season was around 1947. I had one of my babies during that last salmon run. Some of the boys went out to fish one last time. I remember it because when Indians have a new baby, they aren't allowed to have any fish or meat or any kind of fat. So my brothers came in with all these salmon and said, "Oh no, Sister can't eat our fish." I remember my mother making beans for me, just beans; no salt or seasonings are allowed either, so my mother would make me tortillas without any salt too. We also ate like that when we were having our period—no fat at all, no meat, no fish. But when you're hungry, even plain beans taste good and you could eat all the acorn you wanted. But I didn't get to eat any of that last catch of salmon from the San Joaquin River. I missed that.

I abided by the Indian traditions as long as I lived with my mother. That's the way she did things. Of course, as soon as I left home I stopped all that. I wish I knew more to tell you about life as an Indian, but that was a long time ago.

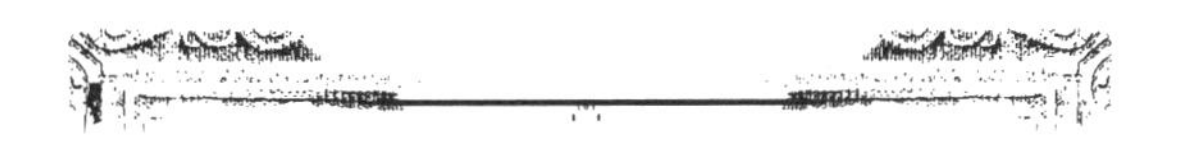

Everard Jones

At eighty-seven years of age, Everard Jones, retired teacher and coach, told his passionate tales of fishing the San Joaquin River. He proudly demonstrated the use of the salmon spear his blacksmith father forged for him many years ago and read aloud the only poem he ever wrote, an ode to his beloved salmon. At his Irish wake a few years later, Everard's family showed this interview DVD to his gathered friends and let his stories live again.

I KNOW EVERARD IS A STRANGE NAME, a hard one to pronounce: Ever-A-R-D. Most people want to call me Everett and it's all the same to me; you can call me whatever you want. My big interest here is in giving the fish some recognition; that's why I wanted to tell you some of my stories.

I would say that maybe 75 percent of the people living in the area now don't even know that salmon ever came up here. The San Joaquin is the second-largest river in California, and the salmon would come through the Golden Gate all the way up here. Sure, some of them would go up into the Tuolumne River and the Merced River, but this was their main river. They would go all the way up to Kerckhoff Dam. Right in these riffles—Rank's Riffle and Cobb's Riffle—is where the salmon would start to spawn in the fall of the year. Of course, that's why they came up the river in the first place. My gosh, the fish! The fish were everywhere and I had the good luck of being right here where a lot of that wonderful fishing was going on.

When there was a salmon run, we first heard about it when the fish got to Gravelly Ford. Someone would have seen a big one. Well, rumors, rumors. But then someone spotted another one. The river was riled with lots of sand particles. It might only be two feet deep and you still couldn't see the bottom of it, but just about the middle of June things started to clear up. Then you'd begin to see salmon. There would be maybe seven or eight in a bunch. Maybe the next bunch would have about twenty. First you'd see their wake off old Lane's Bridge. And wow, what excitement! One of the highlights of my career was spearing two salmon in one day. It happened on a Sunday in 1930, off the old bridge, upstream about a mile from where the new bridge is now. Two salmon in one day!

We originally lived on forty acres at Herndon and Blackstone. It was a pretty good location because it was only about a mile-and-a-half walk to the river. We'd carry our spears down there with us. It's called "spearing" or "gigging"; those were the terms we used. We made a gig out of a pitchfork. My dad ran a blacksmith shop just across the street from the Sportsman's Center. It was about 1898 when he started down there—right

Gigging for salmon. Courtesy of Gene Rose

Everard Jones with the ten-foot plywood fishing boat he purchased for fifty dollars, circa 1943, Courtesy of Cathy Jones

about when he and my mom got married. In those days a person had to be a jack-of-all-trades. Dad could do anything—mark off timber, build a whole house—but he made his money as a blacksmith, with coal, the old-fashioned type of blacksmithing. He'd take a pitchfork and make it into the nicest spear you ever saw. We'd just go across from the Sportsman's Center and we'd spear fish right there. I still have a gigger, I'll show it to you, it's in the back of my pickup. They look like the devil's fork. It would actually be about twelve to fourteen feet tall, because quite often we speared out of a boat or we might wait down by a riffle and spear from the shore. You spear a salmon down the spine. It's not sporting, not a real sporting event, not like catching them with a rod and a line. But they were big devils and a fisherman's got to be careful. You can't imagine what that was like. The Chinook salmon could get to be around fifty pounds. That's a pretty good-sized fish and there were just oodles of them. I'll tell you, it was a lot of excitement.

We'd gig them in the river and carry them back home, but sometimes we would even have trouble *giving* a nice salmon to someone. Years later my wife and I would go down to the store or to a restaurant to get some salmon—very expensive stuff, eight dollars a pound—and I'd say to her, "Lorraine, I remember when we used to have difficulty *giving* a forty-pound fish away!" You know, times have changed. My gosh, back then no one wanted your fish because they could catch 'em themselves, easy.

We didn't save the fish; you'd either eat them right away or get rid of them. The Indians would dry the salmon right there on the river where they caught 'em. But the white men never dried them; we were too lazy, way too lazy. I never actually fished with Indians but I was often on the river when they were fishing. The Indians would wade out on that river with only a one-point spear; that was much harder work than a three-pointer, but boy, they were good at it.

Now, I thought I saw a lot of salmon in my day, but my dad had been here since 1898. He was along the river all that time and he'd say to me, "Sure, it's a good run this year, but this isn't like the big ones." When he was a youngster, he'd drive a horse with a buggy or a buckboard and he'd ford the creek where the riffles were, where the rocks were, because that was the shallowest place to cross. That's also where the salmon spawned. He told me about times when the salmon would cut through the water around the buggy—maybe a dozen fish swarming all around—and it would scare up the team and frighten the horses.

There were some big floods in those days before the dam went in. I guess it was 1937 when it ran right through Clovis. Some people had to evacuate their homes. The water ran all the way down the banks against the Santa Fe railroad tracks down into Fresno. I remember Mrs. Gaines; she had a home around there that was made out of adobe. It just melted. She had nothing but a mud puddle there and they lost everything.

The floods scour the river. The reason there are a lot of willows in the river now is because of the floods. I drifted down there one time after a flood and hardly recognized some of my old fishing holes; it had all changed so much. Before the dam went in, about May, the river would be real high. You'd see half a pine tree coming down, big chunks of wood just scouring the river and ripping out any willows as it went by. When it all settled in the middle of June you'd have a nice sandbar and good fishing.

I remember Everett Rank, Sr. telephoned one day when I was just getting ready to go into the service. He said, "I found a guy who has a little ten-foot boat and a motor and wants to sell it for fifty dollars. Are you interested?" I said, "Oh, boy, yes!" That was a good buy. I used that boat nicely. Everett was quite a fisherman; for years he and I would drift down the river and spear salmon out of that boat.

At some point it became illegal to spear salmon. I have to admit that I was sometimes one of the lawbreakers. Everett and I and others, we would get real bold. We'd go out at night and put a Coleman lantern about midway on the boat. We'd reflect the light so it didn't blind us, but it would light up the water and blind the fish so they were easier to spear. That's not very sporting, really, so I should probably keep that quiet.

The rod that I caught salmon with after we completely stopped gigging was a bass outfit, really for black bass. This was before spinning reels were invented. I would have loved to have had one of those spinning reels, but I used just a little Shakespeare reel. I would wade out on the riffle, cast out, and thumb the line to stop the salmon. I would just dip the line in the water, and if I didn't stop the salmon quick, I lost him right away. Can you imagine catching those big fish with that little pole? If the salmon was caught in a swift current, it could not be stopped. The fish used the current to their advantage; I think it gave them almost double strength. Sometimes there were so many salmon in those riffles that just when one would pull off, while I was there reeling my line back in, I would get another one on the return.

Now the last salmon—and I can vouch for this—their last run up here was 1947. Someone may have planted some little ones around here after that, but when they shut the water off with the dam, the salmon didn't have any water to come upstream with. I don't mind telling you that was a sad day. I learned to fish for trout, but it never was like those days of the salmon.

At one time I knew everybody who lived on the river. Bob Gross's dad lived there for a while and the Dawsons were

"My dad ran a blacksmith shop...It was about 1898 when he started down there....He'd take a pitchfork and make it into the nicest spear you ever saw." —*Everard Jones. Courtesy of Gene Rose*

nearby. The Rices were down farther, but they weren't fishermen. Thank goodness the Cobbs didn't fish; there were enough Cobbs in the county to have depleted all the salmon in the river.

Everett Rank, Sr. lived on the river too; his house is still there. I just loved Everett. He was an awfully nice person, and I honestly believe we would hardly have a drop of water in this river if it hadn't been for old Rank. I know he didn't do it by himself but people didn't flock to him either. He got verbal assistance but that's all. He had to do a lot of fighting by himself but he stuck with it. He finally reached an agreement with the government that a certain amount of water would be designated for the river. We all owe a lot to Everett.

I wrote "The King of the San Joaquin" when I was in the service. I never went overseas but I was down in the Mojave Desert, and in Texas and places like Lancaster, California. It took me about a year and a half to write this poem. I'd never in my life done anything like this—written a poem—but I got so homesick. I yearned to see the fish and here I was out in a desert—in this case Yuma, Arizona. That's about as far from salmon as you can get. It's an amateur's work but I was passionate, I really was.

The King of the San Joaquin

Out of the Pacific and through the Gate, and up
the river to spawn,
comes a gallant knight with plenty of fight
to cope with the dangers beyond.
He carries no scars, this warrior king.
He's king of fish in the San Joaquin.

His hardships are many, his dangers untold, as he journeys
against the stream,
yet onward and onward he battles his way, impossible
that it may seem.
He navigates using no compass or stars.
He's the king of the San Joaquin.

He has many names this warrior bold, from warm
waters to the cold,
Chinook, Quinat, Silverside, Red, he's known quite
well unto all.
He's fat when he starts, but later gets lean,
He's king of fish in the San Joaquin.

They've heard that he's coming, the salmon so fine,
they've rigged up their outfits and tested their line.
They've checked their boats and are waiting to greet
The king of the San Joaquin.

I tried a cast I had tried before;
the current carried it away from shore.
I was winding it in, as I always do,
when the line snaps tight and the rod bends true.
The king of the San Joaquin.

The battle is on and there's plenty of fight.
I thumb the reel with the greatest delight.
The rod is high and he takes out line.
No need to hurry; he's a-biding his time.
He cuts the water so swift and clean,
He's king of the fish of the San Joaquin.

My thumb is burned, my fingers are sore,
I'm ready to quit, but he's asking for more.
When out of the water with a mighty splash,
a glimpse I get, a silver flash,
and back he goes with a terrific thrash,
The king of the San Joaquin.

He finally tires as all fish do.
I thought he had quit, I thought he was through.
I reached for the gaff, started to gloat, as I brought
him alongside the boat.
The king of the San Joaquin.

I was ready to gaff him and end it all there,
when he takes a last jump straight into the air.
It took him a second, a second too soon, to shake
his head and toss a spoon.
The king of the San Joaquin.

I relaxed in the seat and lowered the rod,
I reached for my pipe and for that I thank God.
I draw on the "burner" and smoke fills the air.
I think of that warrior, that salmon so fair,
The king of the San Joaquin.

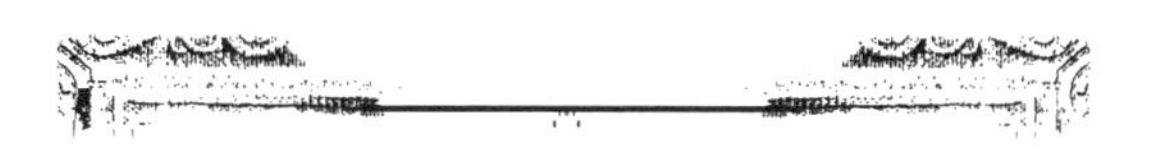

Azalea Ball Biglione

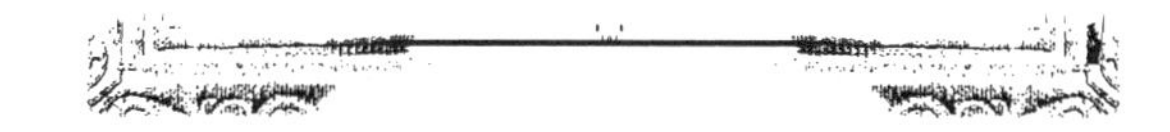

At the age of ninety-four, Azalea Ball Biglione vividly remembered the wonders of growing up on the Ball Ranch, one of the largest and most well-known properties along the San Joaquin River. Though she is now deceased, the stories of her pioneering family and a wonderful lifetime spent along the river live on.

I GREW UP ON THE SAN JOAQUIN RIVER on what they now call the Ball Ranch. My father was Harry Ball and my mother was Pearl Drury. We had owned the ranch for years before we actually moved there; Dad had a house put on the property for some people who worked for him. We lived in Clovis first, but my mother always wanted to get out onto the ranch, as she loved the country. We moved out there in 1916.

We moved in some old houses from the foothills, from up on Watts Valley Road. There was a mine up there with some buildings that had been used for cookhouses. My Grandfather Drury had some moving equipment that he'd brought with him from Missouri. He moved a lot of houses around Clovis at that time but only if the house was worth moving, because it was a costly process.

They used horse or mule teams to pull the houses. They had to jack them up and put them on rollers, and that took a lot of animal power. Grandpa and Dad moved two buildings down to the ranch with a mule team. They were just old shacks, but we put them together and ended up with a fair house. It was a long room, like a dormitory, and they built a bath on one end. There was a sleeping area in the back and a porch; we had screened porches all around the house. We didn't have refrigeration

"Once my grandfather knew that we were coming out to the ranch to live, he offered to move a one-room schoolhouse down so we'd have a school close to us....it was the Millerton School originally."—Azalea Ball Biglione. Courtesy of Eastern Fresno County Historical Society

Traveling from Minarets (north of Friant), this locomotive is pulling a load of sugar pine bound for the Sugar Pine Lumber Company, in Pinedale, California. Courtesy of Eastern Fresno County Historical Society

back then, but we kept fairly cool on those porches. That's what our house was like until we finally moved the front part of the old Burckett house over too. That house had two bedrooms and a big living room, so by then we had a nice big home.

When we first moved to the river, my mother was afraid we might drown in it, so she immediately hired a man who lived above us, Clyde Martin, to teach us to swim. All he did was just throw us in the river, but it sure did work. From then on we swam a lot; we swam the horses in the river and we fished too. When they were old enough, Willis and Ike built a boat with an airplane motor on it, and we learned to ride on a board behind it. We'd go up and down the river, fast. Of course, there was more water in the river at that time, much more than we have now.

We used to fish when the salmon came up the river in the fall. In the spring when the suckers were in Dry Creek, we would catch them with our hands. A sucker fish has lots of bones, so we'd feed them to the hogs. We'd just throw them onto the banks, pick them up in sacks, and cook them for the pigs. We utilized everything back then. We had a lot of chickens too. I hated those chickens, but my mother was always thinking about making money and that's why she got them. She had the coops built and bought a lot of very, very good breeding hatchery chickens. She'd keep all the hens and sell the fryers. That made her more money, and then she started getting roosters for her hens, and she raised hatchery eggs, too.

We also had alfalfa. We could really grow things there. We had a big reservoir behind the house and wonderful wells because the water that came down the river replenished the underground source. They don't have that kind of water now, not with the dam up there.

Once my grandfather knew that we were coming out to the ranch to live, he offered to move a one-room schoolhouse down so we'd have a school close to us. There was an old school up where the Indian mission is now; it was the Millerton School originally. When it was moved to Friant, it was called the Pollasky School. It was moved down to where the White Friant Ranch was—the ranch just south of Friant that has apricot trees on it now. They moved the school building right where those million-dollar houses are being built now. That school stayed there until they moved it to Fort Washington, where it stayed for a long time. It was really a portable school; that's the truth.

I remember one particular time when the river flooded. There were a lot of people during the Depression who lived just south of Friant along the river on the Wishon property. They had come out from the Midwest to pick cotton and fruit. They came to have a better life in California. Well, during that time there was a strong flood that took all those people's tents and everything they had—right down the river. But they came right back and settled along the river again. Some of them were still looking for gold at that time too. No one ever bothered them because they didn't disturb our property and the gold was in the water, so it was legal to pan for it. Later things went downhill a little when they started stealing our calves; finally we had to get the help of the sheriff's department and move them out. By then it was just riffraff up there; most of the original settlers had found homes. The sheriff's department moved the rest of the people to places where they could live better lives, and they all settled in the area.

I remember the old railroad too. It was still in operation for years after we moved to the river. The passenger train ran twice a day, morning and night up to Friant from Fresno. At that time they were freighting things up to Auberry and Big Creek when the dams were going in. Later they used that same railroad, the Sugar Pine, when they built new rails to bring the big trees down from the mountains. In fact, when I started high school, there were kids going on the train to Clovis High every day; sometimes I would go on the train too. We waited at the station in Friant and the train stopped to pick us up. It was a long ride!

We lived near a deer trail. A lot of those deer were little fawns that hikers or hunters had picked up in the high country. They were orphaned because their mother wouldn't take them back after they'd been touched. Those poor fawns would end up at Roeding Park in Fresno. Billy Holmes kept an eye on wildlife there. He was a friend of Mom and Dad and he came out to our ranch all the time. When young animals were taken to the park, he'd bring them out to us because we had a dairy. Us kids would feed the deer with our calves and when they were big enough to turn out, we would just put them out into the fields. Many mornings when I'd ride out to check the cattle, there'd be deer right in with the cows. I expect some of the deer on the river right now are related to those orphaned fawns.

About three years ago, I went down to the ranch. I'm ninety-two now, but I'm still driving and I get around just fine. So I went to a place where I knew the deer went through. I drove my pickup down there, took a book, and sat in the shade of a tree and read and watched the deer go down their trails. I saw a mountain lion that day; he was waiting for a deer to come by, so I drove up and scared him away, but he probably only went to another tree someplace. When you live like we did, you enjoy seeing things like that every day. It's not unusual; it's your life.

FRESNO AND BAKERSFIELD SUBDIVISION. 9

Eastward	FROM SAN FRANCISCO						TOW. S. FRAN.		Westward
	THIRD CLASS	FIRST CLASS			Time Table No. 136 February 27, 1921.		FIRST CLASS		THIRD CLASS
Capacity of sidings in car lengths and location of Scales, Fuel, Water, Turning and Telephone Stations.	320 Way Freight	148 Fresno Friant Passenger	146 Fresno Friant Passenger	Distance from San Francisco		Distance from Friant	147 Friant Fresno Passenger	149 Friant Fresno Passenger	321 Way Freight
	Leave Daily Ex. Sunday	Leave Daily	Leave Daily		STATIONS		Arrive Daily	Arrive Daily	Arrive Daily Ex. Sunday
YardWFTOPY	6.30AM	4.00PM	8.00AM	205.5	DN-R FRESNO 1.6	24.4	s 10.55AM	s 6.55PM	12.30PM
I				207.1	A. T. & S. F. CROSSING 0.9	22.8			
				208.0	EAST FRESNO 1.4	21.9			
	6.50	f 4.15	f 8.15	209.4	BARTON (Spur) 2.2	20.5	f 10.41	f 6.41	12.05PM
				211.6	GRANZ (Spur) 0.2	18.3			
	7.00	f 4.21	f 8.21	211.8	MALTERMORO (Spur) 0.3	18.1	f 10.35	f 6.35	11.55AM
				212.1	NAVIN 0.8	17.8			
	7.05	f 4.24	f 8.24	212.9	LAS PALMAS 0.3	17.0	f 10.32	f 6.32	11.45
				213.2	FRESNO INTERURBAN RY. CROSSING 0.4	16.7			
	7.11	f 4.27	f 8.27	213.6	EGGERS (Spur) 0.3	16.3	f 10.29	f 6.29	11.40
				213.9	VANRIS 1.0	16.0			
	7.17	f 4.32	f 8.32	214.9	TARPEY 1.2	15.0	f 10.24	f 6.24	11.30
	7.23	f 4.36	f 8.36	216.1	MELVIN 1.3	13.8	f 10.20	f 6.20	11.20
W	7.40	s 4.40	s 8.40	217.4	D CLOVIS 1.1	12.5	s 10.16	s 6.16	11.12
	7.48	f 4.44	f 8.44	218.5	GLORIETTA 2.4	11.4	f 10.12	f 6.12	10.45
				220.9	SETCH 2.0	9.0			
25	8.06	f 4.58	f 8.58	222.9	GORDON 0.7	7.0	f 9.58	f 5.58	10.25
	8.15	s 5.01	s 9.01	223.6	EL PRADO 1.2	6.8	s 9.55	s 5.55	10.20
		f	f	224.8	BURKHEAD 5.1	5.1	f	f	
T	8.45AM	s 5.25PM	s 9.25AM	229.9	D-R FRIANT	0.0	9.30AM	5.30PM	9.45AM
	Arrive Daily Ex. Sunday	Arrive Daily	Arrive Daily		(24.4)		Leave Daily	Leave Daily	Leave Daily Ex. Sunday
	(2.15)	(1.25)	(1.25)		Time over District		(1.25)	(1.25)	(2.45)
	10.84	17.22	17.22		Average speed per hour		17.22	17.22	8.87

"I remember the old railroad too.... The passenger train ran twice a day, morning and night up to Friant from Fresno."—Azalea Ball Biglione. Courtesy of Fresno County Library

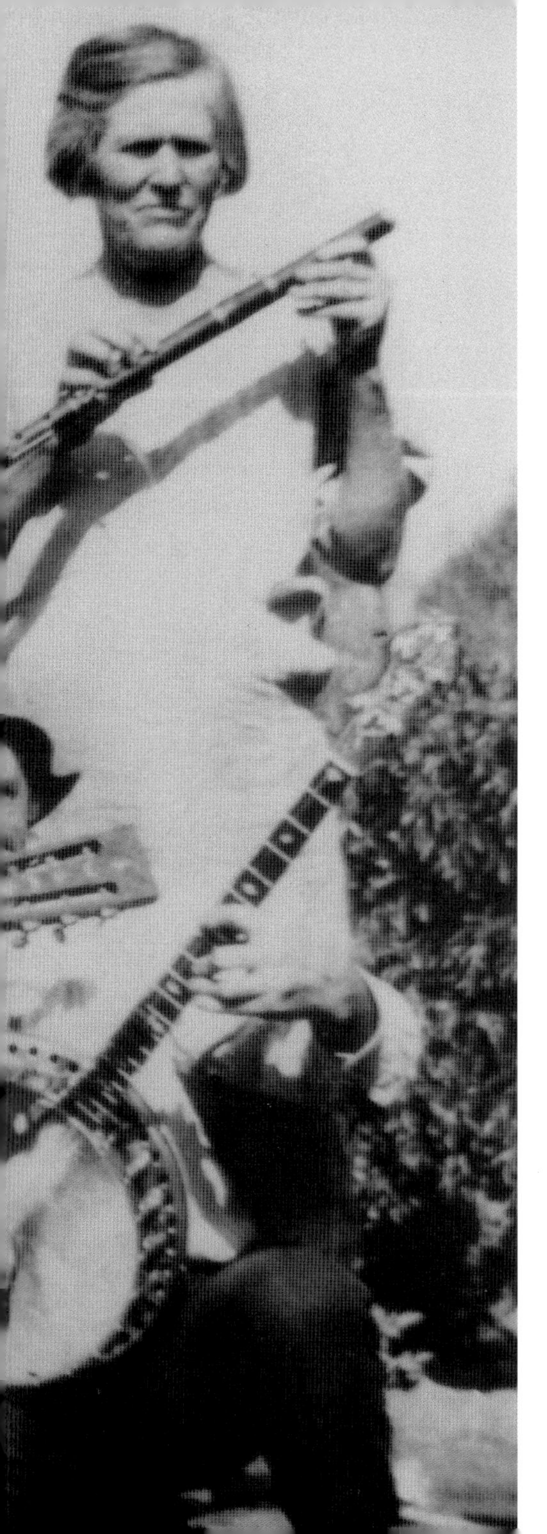

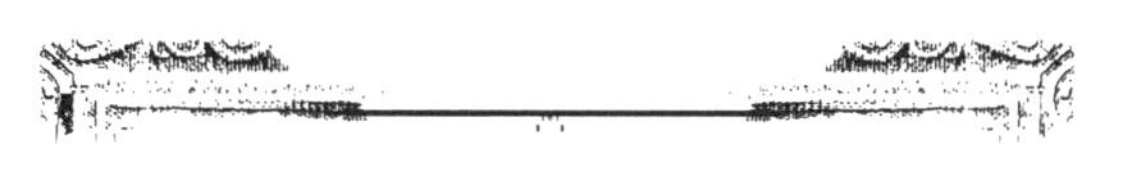

Viola Adlesh

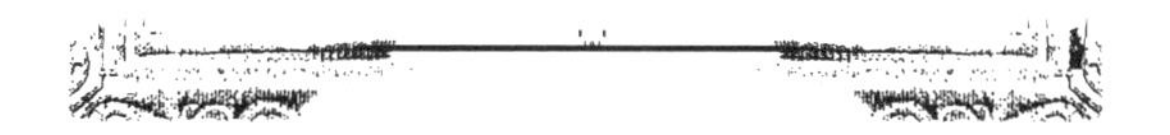

Viola Adlesh still lives on the rugged land her gold-mining family homesteaded at Temperance Flat. From her deck overlooking the river she shared her family photographs and a lifetime of memories on the San Joaquin.

IT WAS 1919 WHEN I MOVED OUT HERE by the river. I was only six months old. My grandparents homesteaded this property sometime in the 1890s. I'm not really a hundred percent sure when, but I have it all written down somewhere. They came up to mine gold. My grandfather had died earlier in 1919 and my grandmother was left raising their children, seven of them, three girls and four boys. They were all teenagers or older by that time. Then my mother died that year too, so Grandma took me and my two brothers and did most of our raising too.

Most of the gold mining happened before I was born, but I heard the stories. There was a lot of mining up here on every little creek—even that little one right down below this house. If you look, you can see how the rocks are all still stacked where the miners put them. They mined from the creeks and out of the ground—anywhere they could find gold. A lot of the tunnels are still around. They go straight back, really deep; some are probably a quarter-mile long. There were so many people up here back then that they supported a little store, right down the hill from here. The customers were all miners. There was a lot of gold back then, but a lot of people were after it.

I do remember some stuff about the gold. You know where that gravel pit is, down on Friant Road? Well, Grandma found gold nuggets there. When she'd find gold, she'd process it and make it into a little ball. I don't know exactly how it worked, but my grandma would have it in the palm of her hand, this big thing of

A musical family. Courtesy of Sally Adlesh

Viola Adlesh with her children Sally and Joe, circa 1945. Courtesy of Sally Adlesh

gold, and we'd cash it in down in Friant. Then we'd buy peaches; there were some big orchards around there and she'd can them so we'd have peaches all year round.

When my aunts and uncles were young the family only had horses and wagons; that's how they made their trips to town. They had teachers come live up here with the family; they were wealthy enough for that then. While they still had their gold money, my grandparents always had the latest thing—Model T's and all. They did real well gold mining, but they put all of their money in a bank, and that's when the banks went broke. Of course, you've heard about that. Anyway, that's what happened to their gold money. My family kept mining right up until the late 1920s, right up until the crash, but that was bad, real bad. They got a little of their money back, eventually, but it was mostly gone. Gone. That would be something now, if banks could still go broke. You have money there and then you don't have anything at all. Wouldn't that be something?

It was a great place to grow up. We swam in the river all the time. It was ice cold. And boy, it was a rushing river before the dam went in! There was a nice round pool that we kids could swim in. When the river slowed down in September and October, there were places where you could walk across it on the rocks and boulders. In the summer, Grandma would do her wash on the river and throw the clothes over the willow bushes to dry and then drag it all back home. They cut those willows down when they made the lake. This house is about seventy-five feet above the high-water line when the lake is full, but when the lake is low it looks like it did back then; the river is still so beautiful.

My Aunt Flora and I were the cattlemen of the family. The cattle would get way up by Table Mountain and we'd have to go around and shag 'em back. We'd be out on our horse looking for the cows, and we'd go up to Big Sandy and stop to get a drink out of those big caves. The water flows underground there. It comes down and makes big potholes; there was always a dipper on the rock that we'd use to drink that cool water. It was so good. It was a soft water, sweet, so different from any other kind of water. I don't know if you'd even drink water from the river now when it's running—probably not.

We had a herd of goats that went up to the top of Table Mountain all the time. You'd look up there and on the tip you could see these little tiny white spots, our goats. I was a teenager and it was my job to go up there and run them down. I'd go all the way to the top of Table Mountain to shag the goats before I went to school. They'd come home with me because they knew they'd get something good to eat. There's a big pile of rocks where we had a pen that they'd come into for the barley. They'd run around on that rock pile and one of them would go up way on top of the rocks. One of my uncles would shoot it; it would fall down and my uncles would dress it out. They'd put the heart and liver in a bucket of water and run home with the goat. We'd have it for

"My uncles always carried guns with them. They always had a .22 when they hiked up to the mines to work. They'd shoot food for the table: a rabbit, a quail, a dove, or anything that was edible. It was good food." *—Viola Adlesh. Courtesy of Sally Adlesh*

dinner. We didn't have refrigeration but Grandma could fit half a goat in the oven—the kids and the small goats. We sure liked that goat meat; Grandma really fixed it good.

My uncles always carried guns with them. They always had a .22 when they hiked up to the mines to work. They'd shoot food for the table: a rabbit, a quail, a dove, or anything that was edible. It was good food.

I remember those salmon—oh yeah, they were wonderful. You could stand on the bridge that came across the river, the one that is right below where the dam is now, and the salmon were so thick it looked like you could walk across their backs. They would spawn right up here, just a little farther up the river. My uncles would spear them with these long poles. All of the salmon would come to this one narrow spot, a place called Gus Smith Falls, some waterfalls named after a famous gold miner. The boys would tie ropes around themselves so they could lean way over and spear the salmon. I think it was against the law, actually. I think only the Indians were allowed to do that, but anyhow, they'd come back with these big old salmon, as long as I was tall at the time. It was exciting but it didn't last too long. They built the dam and did away with all that.

My grandma lived to be ninety; she was always a tough lady, right up to the end. When fishermen wanted to come down our road to fish here, Grandma would send her grandchildren to stand at the road and collect fifty cents from them. She made them pay to fish on her property. Good for her!

There was a huge tree right outside the house and it had the most delicious figs. For years, the Indians would come up to trade salmon for those figs and the deer would come around to eat them too. Even after Aunt Flora was living there all by herself, after everybody else had died, she'd sit out there under the fig tree with her gun and get herself a deer every season. We lived under that tree all summer. They were just little figs, but they were so good! They are all gone now. The fig tree is completely dead. I tried several times to get slips off that tree and start another one but—no luck—I couldn't make them grow. But we still have wild grapes down by the creek and blackberries on the river—if we can beat the deer to them. The bears eat the grapes, too. I don't remember if we had bears here when I was a child but these last years we've had quite a few. There's still a lot of wildlife.

We've always been pretty isolated up here. One day Uncle John had been fishing and he was walking home from the river up where the old barn was. He always carried a pistol in his belt and he somehow stumbled and fell. The pistol hit him in the head, discharged, and killed him instantly. I remember exactly the moment I heard the news, because he died on the day I was delivering my youngest child. That made it especially hard for me. Anyway, Aunt Flora was the only one at the house, so when Uncle John didn't come in by dark and she'd heard that shot earlier, she went looking for him. When she finally found him, she couldn't really do anything about it until the next day. For one thing, she didn't drive. She was in her seventies or eighties, but she had to ride her horse out to report Uncle John's death.

When we first got married, my husband hated it up here. He was a city guy. But when I was given this little piece of property, seven acres, I said, "I'm moving back up there." He said, "Well, I guess I'll have to go too."

One evening last summer I was sitting in my rocking chair, and here's this darn rattlesnake right at my feet—in my living room! This one was sort of lethargic, just stretched out there on the floor. I tried to put him in a paper bag so I could dump him outside, but he wouldn't go for that. I ended up killing him with a shovel. Since then I'm always looking down, even in the house. It's pretty rugged living, but I couldn't live anyplace else. It's like I live on my own little island.

Clayton Vander Dussen

Clayton Vander Dussen lives just two miles from the San Joaquin River on a cattle ranch along Cottonwood Creek, the last tributary to the river below the Friant Dam. Now in his eighties, Clayton can still often be seen on his tractor clearing away dry grass and plowing his family orchard. He met his current wife in a watercolor class and they paint together at their dining room table.

I MARRIED MY FIRST WIFE, Dorothy Wagner, in 1948, and my father-in-law gave us eighty acres in Friant as a wedding present. I built the house myself and dug the well. Whatever had to be done, we did it. It ain't modern, I'll say that, but it's livable. We raised turkeys and cows and here we are. It's been a pleasant life.

There's not too much history written about the Wagners but they've been here a long time, longer than almost anybody. Everything I know is from the family stories that get told. The Wagners came here in the 1870s, and at one time they had a store in Friant. But there was a big flood one year—I don't know exactly when—and they lost their store and their house. I know that George Wagner was quite a colorful guy. Apparently he was pretty successful in finding gold. He did a lot of mining. He had some Chinese men working for him, and most of them made a lot of money. They would put nuggets in their mouths and hide away the gold at night. When they got enough, they'd go back to China.

With his store and the gold, George Wagner was able to buy quite a bit of land all around the area. Then one year he mortgaged his sheep for ten dollars and when the price went down to fifty cents he went bankrupt. That's when he lost most of his land. The only piece left was the forty acres that his wife had homesteaded. Back then, if you homesteaded your land, even if you went bankrupt, they couldn't take it from you. After the bankruptcy, the Wagners paid for all their foodstuffs by cutting firewood. That's what a lot of people up here did to make money, and that's why there are not very many oak trees left. It was all cut for firewood for people all over Fresno; they used the wood for heating their stoves.

I remember the night the Friant Bridge went out. I might get into trouble with the Bureau of Reclamation for this, but there was an island on the north side of the bridge and they'd dammed it off. That forced all the water to one side; it washed the sand up into the pilings and that's why the bridge collapsed. We had a friend who was going up to buy sheep that day and he almost ran right into that first hole. My wife and I walked out onto the remaining bridge and looked down; the next day the rest

"There's not too much history written about the Wagners but they've been here a long time, longer than almost anybody." *—Clayton Vander Dussen. Courtesy of Brad Polzin*

of the bridge collapsed, right where we'd been standing. It could have collapsed while we were on it. When Madera County went to clean up their part of the broken bridge, they set off a blast and broke all the windows in Friant. They quit then and never did try to clean up the damage again. It would have been nicer if they had, but I don't think it's ever going to happen.

There is a good view of Lane's Bridge in my photos. Across the river from the bridge was a big sandbar; everybody went to the beach there. I wasn't there the day that bridge fell down, but I remember one time when a herd of cattle overloaded it, and one time a tractor with a heavy load broke it down.

I've got some good pictures taken during construction of the dam, but I remember when I was a kid we used to go up to Millerton before it was a lake. There was a ranch there that McKenzie owned at that time; he owned the whole town on both sides of the river. But Millerton was abandoned by the time I can remember. The town of Friant was there, of course, but it was called Hamptonville first; then it became Pollasky. It finally became Friant because it was White and Friant who put the railroad in that came from Pinedale.

When the salmon were rolling in the rocks and laying their eggs in the river, you could hear them clanging and banging on the cobblestones. They like to be in a riffle because they have to dig holes for their eggs. There's a riffle down by what they called Ulysses Place where at night you could really hear those salmon in the rocks. Some of the salmon were damaged; they'd grown fungus and turned spotted and white. On the trip up they moved from the salt water to fresh and sometimes they'd get hurt along the way. The salmon even used to come up Cottonwood Creek, the stream that runs through our property; it's a tributary to the San Joaquin. In fact, not too long ago UNESCO designated Cottonwood as biosphere reserve.

Everybody went out at night to fish for salmon because it was less risky; the authorities really frowned on spearing. We'd put our boats in at Roulard's Riffle. We had lanterns on the boat made from empty five-gallon tins of gas. One side was cut out for a reflector and the other side was left on so the light wouldn't get in our eyes.

Many of the people who went out at night tried to go over the rock dams, and a bunch of them drowned. At night there was no visibility, and there were some tough spots to get over. I went over them several times, but it was always dangerous. Lots of people drowned in the San Joaquin back then; it was a big river. I remember old John Jones, Everard's father. He had a blacksmith shop just below Lane's Bridge, and his brother Leonard had some property down toward where the streetcar used to end. One time there was a fire across the river and Leonard tried to swim his horse through the river in high water. He got caught in an old willow tree and drowned. People mostly get in trouble when they try to swim upstream. If you go with the flow you will probably be all right; people who fight the current are the ones who run into problems. Just like life.

J.R. McDONALD.

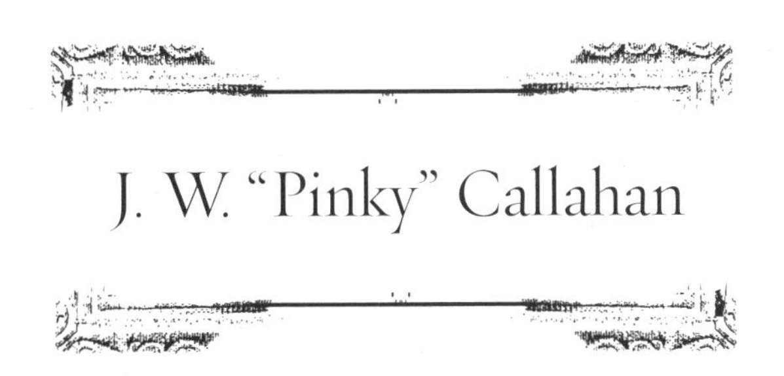

J. W. "Pinky" Callahan

Pinky Callahan was a cowhand and jack-of-all-trades, working up and down the San Joaquin River all his life. In his small house trailer situated scenically in the middle of a large grain field on the Jensen ranch off of Highway 41, Pinky told his story in the comfort of a well-worn lounge chair. He died less than a year after our interview, leaving behind a lifetime of river stories.

I'VE LIVED PERMANENTLY ON THE Jensen ranch for seven or eight years now, but I've been in and out of here ever since I was fifteen years old. I helped Jensen's daddy with the cattle; I also cut hogs and calves for him. He had over two thousand acres at one point; his land went all the way up to Bates Station. He hired me to help him with all his livestock. I knew all of the families along here; I've been working up and down this river forever.

I used to have a dandy horse that I used for my cowboy work. An old guy, Fred High, had traded three or four pack horses for one good one. He didn't know why he bought it; it wasn't broke at all, but it was a beautiful light palomino mare. She had a chest on her, and nice wide front legs. Fred told me, "Pinky," he says, "I got something I want you to come and look at." I'd worked for him. He had thirty or forty acres out on Herndon Avenue; we'd moved a house in on part of it and put in a permanent pasture. He says, "I got something I think you'd like to have. Why don't you come out and look at it?"

I went out there and saw this beautiful young mare—she was about four or five years old. I loved her look; she was chunky and short. I said, "What'll you take for her, Fred?"

He said, "Give me $125 and I'll throw in the saddle."

Two or three days later, before I'd had a chance to pick her up, Fred went out to the pasture and found her down on the ground. He calls me up and says, "Pinky, I can't sell you a crippled horse."

I said, "What's the matter with her?"

He says, "She fell down on the ground and couldn't get up. The vet said he couldn't do anything for her."

So I went out there. "Fred," I said, "there is nothing wrong with that horse; she's just too fat and it's pushing her bones out of their sockets. Just don't feed her for a few days; starve her a little bit." So he did it. He got her off the groceries and it worked. In a week she was running around. She turned out real good. I brought her out here on the ranch, broke her, and we did a lot of good work together for years.

I've been in Fresno since I was about three years old, but I was born in Rutherford, in Napa County. My dad got a job

"I'd hear those old fellas talking about the ferries that used to come all the way up the river." J. W. "Pinky" Callahan. Courtesy of Fresno Historical Society

with the Edison Company at Big Creek, so we came to Fresno on the train. We stayed overnight at a hotel, and the next day we caught another train up to Auberry. They had to put out a snowplow for us to get all the way up to Big Creek. In those days there was snow from the eaves of the house clear to the ground, like popsicles.

My dad was like me, always trying to dig up a better paycheck, so he bought a tree from the Edison Company, a big sugar pine that was over twelve feet through the trunk. He kept sawing at it year-round for the whole time we were up there. He sold that firewood to everybody in the mountains, and four years later there was still half of that tree left. It was quite a production. Dad had a little putt-putt motor and a drag saw. He weighted the upper end of the tree and hooked it up like a crankshaft. Once, we cut off a chunk and it dropped to the ground, came loose and ran over the stopper my dad had set. It bounced about a quarter of a mile to Big Creek Road, jumped the road, and went all the way into the powerhouse. That was something to see.

It was about 1923 when we moved back to Fresno to where the Tower District is now. It was mostly vineyards and orchards then, with a few farmhouses here and there. There were still a lot of horses and buggies then, and it was nothing to see a horse get away from somebody. They would get spooked and run down the road with a buggy and nobody in it. The horse would fall down, get tangled up, and tear the buggies all apart—right in the Tower District.

As kids we used to go out to the San Joaquin River all the time. We'd swim and fish. We'd walk out there from Fresno or sometimes my older brother would take me out in his Model T Ford touring car. I think my brother paid four or five dollars for that car. It quit running and the guy who owned it didn't know how to get it going again. The old guys didn't know how to repair cars, but us boys could figure it out. My brother and I fixed up that old car and then we'd go driving clear up to the old fort, to Fort Miller. We'd go along the river where the railroad track was. There was a little dirt road you could drive up; it was rough, but we had twenty-one-inch tires, so you could go over big rocks without scraping the bottom of the car. We used to go up and muck around the fort. Oh, it was quite a deal.

The town of Millerton was still standing and there were a couple of big buildings there. There used to be a schoolhouse behind Table Mountain. The Scullaries lived back up there in between the mountain and the school. They raised goats. They had one billy goat that was famous along the river. That billy goat stood real tall and he smelled to high heaven; you could smell him from five miles away. He had a beard hanging down to the ground and he'd stick his head between his legs, pee all over the beard, shake his head, and scatter it everywhere. He was one crazy goat. The Joneses had a house with a big barn and corrals on the corner where Woodward Park is now. One day Herb Jones was down on the river with his cattle and mules, and they were all going crazy. Herb ran out, and here comes that goat across his pasture, coming up from the river. Herb and his kids got their horses saddled up, went out and finally got a rope on him. The rope stunk so bad that later they had to bury the darn thing. They put the goat in a hog wire pen behind the barn. He stayed in there about ten minutes, until finally he got a running start and ran up on the fence brace, jumped over it, and came back across the river. He got into the Jensens' cattle. Old man Jensen saw that goat go through the pasture. He got me to try to help, but when that goat got going there was no stopping him. I never saw a goat like that before or since. Everybody talked about him for years.

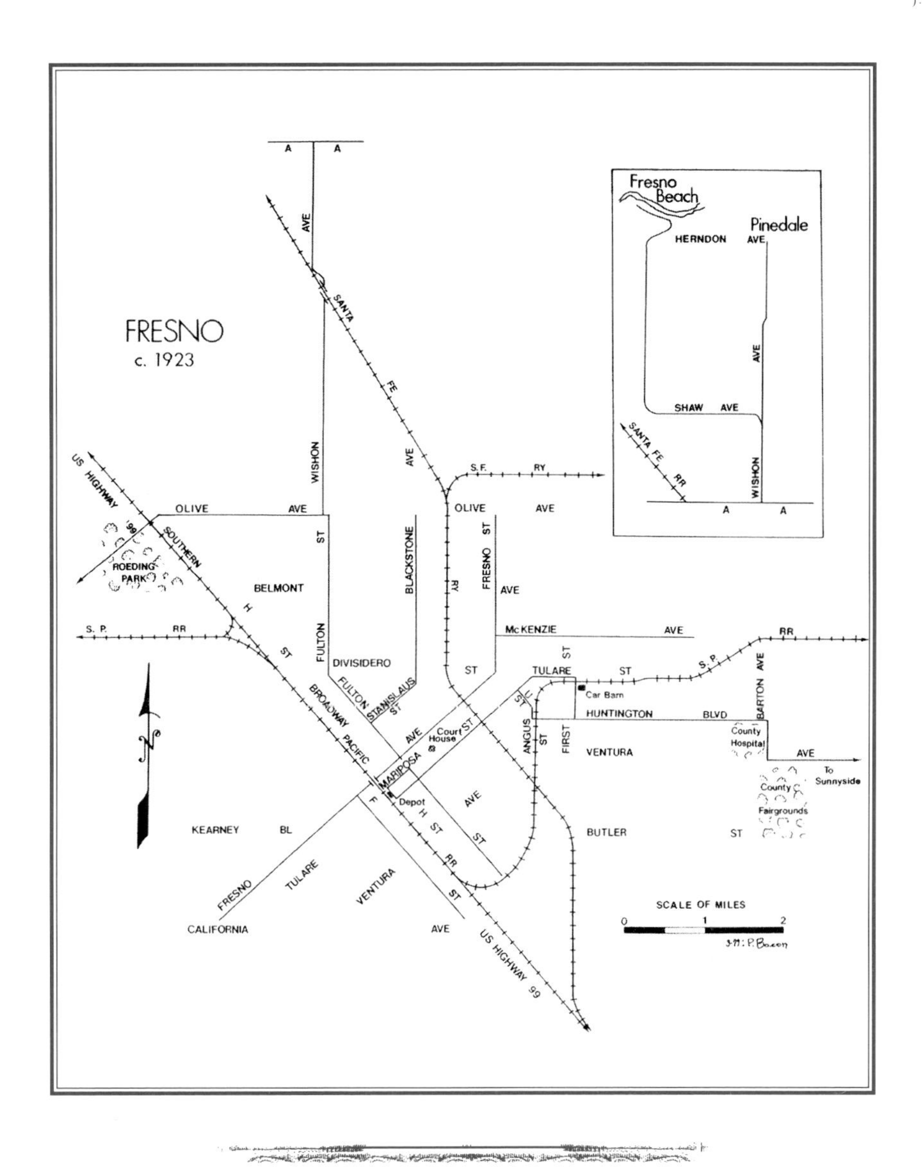

"People used to get out to the river on the streetcar. It came clear into Pinedale."—*J. W. "Pinky" Callahan. Trolley line map courtesy of Fresno County Library.*

People used to get out to the river on the streetcar. It came clear into Pinedale. You'd get off at Herndon; that's as far as it went. I'd see kids head out that way all the time and I knew that's where they were going, to take a dip in the summertime. Nobody had swimming pools then. One day a couple of us were riding back from working at the old Profey ranch. We heard some females screaming. We were way up on top of the hill there. I heard this screaming but I couldn't see what it was all about. So we dropped our horses to the next level and rode up there to make sure there wasn't any trouble. We looked over the edge and saw about six or seven girls. They had a rope tied in the tree with knots on it so they could hold onto it better. They didn't have—well—they had all their clothes off. Not a stitch of clothes. They'd hold onto that rope and swing way out over the river and drop into a hole. They were just a bunch of college girls having a ball, but none of them had a stitch of clothes on. We got out of there real quick.

I got another story. One time the river was coming up high; it was before they put the dam in. Everett Rank had cattle out on his island. He went over and tried to get all the cattle off the island before it got flooded, but there was a jersey cow that had a calf only about three or four days old. Everett called me, wondering if I'd go save that cow and calf. I found the calf right away but I didn't find the mama for a while. I finally found her way out in the brush and she took off running ahead of me on the trail. I followed her around. She wouldn't cross the river, but she ran around and around the island. Every time she looked back I'd be right behind her. I made her run around there about a dozen times, then I ran back the other way. I knew that if she stuck her head out into the clearing I could pop a rope on her. So I ran back right quick and sure enough, here she comes out in the open. I got a rope on her and dragged her across the river. My horse had to swim a little bit but it wasn't too deep. I put the cow in the corral, then I came back, picked up the calf, put it up on the saddle with me and brought it across the river. I think Everett gave me five or ten dollars for that. In those days that was a lot of money.

Things have changed; that's the truth. There used to be more foxes on the river, more beavers, more fish—more fishermen, too. I was fishing down there at Lost Lake not too long ago and I got to talking to a guy I know. I told him that a friend we both used to fish with all the time had died—my best fishing buddy, Ray Badger. All the old-timers are dying. There were ten kids in my family; now there's only me and my youngest sister.

I remember what it was like to listen to the old stories. I'd hear those old fellas talking about the ferries that used to come all the way up the river, but they put that dam in at Mendota and then the ferries couldn't come up anymore. The salmon quit coming; they haven't been up here in years. Those land grabbers wanted water. Now they is hollering they want more water. Seems that's what they're still fighting about today. It's quite a mess, but there's nothing for me to do about it. I'm retired from it all now. I'm taking it easy.

Thelma Russell

At the age of ninety-six, Thelma Russell opened her photo album to show a faded snapshot of herself, her husband, and two friends holding up the salmon they caught on the last legal day to fish in 1947. She shared this story and the stories of her childhood on the river in Kerman from her current home in rural Madera County.

I'VE LIVED ALL AROUND THE VALLEY. We lived in Fresno at one time; there were quite a number of Basque people living there. We were Spanish-Castilian. I was born in Spain. I came here on a boat when I was three years old. My dad came first to work here with his father and brother. My father's uncle had already come about five years after the gold rush, and then his brother came over with his sons in 1910.

We moved from Fresno to Kerman, up on the last road along the river. It was way out there and I tell you, those were hard times then! My uncle had a little dairy on the river's edge, and my dad brought in the few cows he had. The farm was down on the river; in fact, when the river flowed high, the cows were right out there in the water.

Dad found an old house up on the hill that had been abandoned. It was just some boards put up, like the old hillbillies' homes. My dad was making a dollar a day and trying to make a living on that, but my mother knew how to use her head. We had a few chickens and some pigeons, and my dad built a coop high enough so the cats wouldn't get them. We had to climb a ladder to get the squabs; we ate them to keep going. I'll always remember that because it was during World War I and we kids used to say a little rhyme: *Kaiser Bill went up the hill.* We didn't know what it really meant but World War I was a big deal, that we knew.

From our house we could see the river, and every once in a while we saw a houseboat chugging up. This was probably about 1920 or 1921. People said maybe the boats were from Stockton. I remember there were a lot of turtles in the river too—not very big ones, but a lot of them. I remember the boys would flip them over and I'd beg them not to turn them upside down. My brothers would do those kinds of things, but my sister and I couldn't watch. You know how little boys are.

We had to walk to school from our house. It was about four miles. Up the hill from us there was a French family, and we had made friends with them. The mother was real sweet and she knew we didn't have anything, so she'd give us chocolate milk. The family had a dairy too, and the mother would fill an empty lard bucket with chocolate milk for the four of us. She'd

Thelma Russell, age seventeen, atop the hay wagon of the family farm. Courtesy of Thelma Russell

have her boys carry it up to our house and she'd send her oldest boy, the smartest one, to come looking for us at school to make sure we had something to eat. He'd bring us food, maybe a piece of bread, because she knew we were hungry. I've been friends with that family ever since. The school was called the Sunset School and it was just a one-room school. I have a picture of it. I think about that family now, as most of them are passing away. They were good people.

My husband and I used to go through Millerton, before the dam. In fact, my husband didn't like the idea of building the dam at all. He was a fisherman and right up above the dam is where the salmon used to spawn. When they put the dam in, he couldn't go salmon fishing anymore and the river sure got quiet after that.

I haven't always lived on the river. I've lived all around, but I'll tell you a story about when I moved out here again with my husband. One year they said we could spear salmon legally. I don't like to fish and didn't like the idea when my husband said, "Let's go spearing!" You were allowed two salmon each, so I had to go if we wanted to get more fish. We all went down there by Los Banos where the river comes in, right before you get into town. There were a lot of people, since everybody wanted to get a chance to spear again. It's the first experience I ever had salmon fishing. Our good friends from Biola made me a spear. It had a long pipe and you threw it at the fish. I got in the water but I was scared to death. We were in the water up to our knees. The other guys got their limit before I did, so they couldn't spear anymore. The salmon were coming upstream and they told me to spear the fish before they got to me. The spear had a string tied to the pole so the fish wouldn't get away. You had to throw it hard and strong enough to make it stick. I got one. That was quite an experience, it really was. That's a time I really remember because spearing a salmon was quite an achievement for me.

I'm starting to forget things, but people tell me I still have a good memory. I have an older sister and she has trouble remembering some of the people I just told you about. People have asked me about getting old. One man said, "I would like to know what you do to keep living so long." I told him, "Don't drink, don't smoke, and don't chase women." That's what I told this guy and he believed me, I think.

Thelma Russell (second from left) with her limit of two salmon, circa 1945. Courtesy of Thelma Russell

"I went to the Millerton School, right there at Table Mountain....But around 1945 I helped tear my own school down."—*Lewis Barnes. Courtesy of Eastern Fresno County Historical Society*

Lewis Barnes

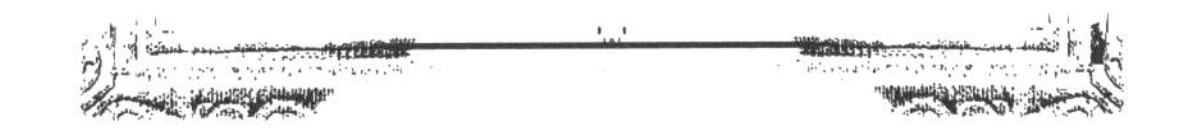

Lewis Barnes was one of the original founders of Table Mountain Bingo, which evolved into the very successful Table Mountain Casino. Lewis grew up on tribal land near the San Joaquin River and spent many happy hours fishing and playing there. He served several terms as tribal chair of the Table Mountain Rancheria and now serves on the Tribal Elders Council.

I was born in 1933. I grew up on tribal land at Table Mountain. We kids liked it because we had the free run of the whole mountain. It was good then. It's good now. I've lived there all my life.

I never learned the old language. My mom didn't either. Well, I guess she spoke it for a while but then they sent her to the Sherman Institute in Los Angeles; it was an Indian school in Riverside. Every time the kids would start to speak their own language they'd tape their mouths or slap their hands or send them into the corner. Quite a few of my friends went there but I didn't have to go. My mother was just a little girl when they sent her down there and the students weren't treated very well, not at all. It was bad. My mother didn't want me to have that same experience, so I got to stay here.

My dad was from the Tollhouse area. He never told me many stories. I really don't remember my mother or father telling any stories at all; I think they were busy doing other things. But I do remember many things about them. My dad and uncles would fish in the riffles where the dam is now. They used to mine for gold there, too, and my father had a mining tunnel where the fish hatchery is now. They'd take whatever gold they could find—not much—but enough to help support the family. That was in the late 1930s during the Depression, so any little thing helped.

We were fruit pickers. I remember that we worked for Mr. Joe Moore in Clovis; he had about eighty acres out there on Fowler. We took care of it for him through the harvest season. It wasn't so bad; what I hated most was picking up the raisins and stacking the wooden trays when it started raining. Lizards would run up your pant legs. I remember that.

I went to the Millerton School, right there at Table Mountain. The school was on Sky Harbor Road. There was probably room for about twenty children—one room for all eight classes. But around 1945, I helped tear my own school down. They said there weren't enough kids to keep it going, so they got the older boys to take it all apart, and they had me rip the nails out to keep anybody from stepping on them. They paid us to do that, to pull our own school down. Come the fall season, they had to send us to school in Auberry on buses. It was a long ride up there and back every day.

The San Joaquin River was very important to us. Every year, in the fall, we'd set up our salmon camp on the banks of the river. Mr. Rank owned all that property along there. He had a road that went all the way back to the river, and we had permission to use it. There must have been eight or ten families that would go there every year. We camped together, everybody we knew. The Walker family, the Smith family, and I think the Tex family from North Fork. I think the Pomonas and the Samples from Auberry were there too. We fished with spears at night with a big basket made out of chicken wire that we used for light. We'd put the basket at the end of a pole and light it; then somebody would have to go out into the water and hold the lantern up. The salmon would come into that light and then the guys would spear them. I was holding the pole one night and a big salmon came by and hit my legs and knocked me down. I fell down and the fire went out. I remember that like it was yesterday.

While the men went to work during the day, picking fruit or whatever, the ladies set up lines to dry the salmon on. They would clean up the salmon heads and boil them and that's what we would eat for lunch sometimes, the meat from the heads. We had a lot of salmon; every day there'd be fish to eat. Everybody would get all the salmon they needed.

As kids we'd try to go see what was happening while the dam was going up. They wouldn't let anybody all the way down in there, but of course we kids would sneak on to the high points on the hills and look down and watch. I remember playing around the buildings that were left from the town of Millerton, but it didn't seem like a town then because they were in the process of cleaning everything away and getting ready to make the lake. As the dam was built and the lake started coming up, I remember we went out in the winter and picked from the orange trees that were left in the old town. There were lemons and oranges. We picked those while we still could. Then the dam was built and the lake started coming up. Another stage.

I didn't graduate from high school. In 1950 the Korean War had just started, so I joined the service. I'd just turned seventeen and I went home and told my mom I was leaving town. She said, "Where you going?" I said, "I'm going to Korea." She didn't like that, but she had to go down and sign the papers for me. She signed them but she didn't want to.

I came back home to Table Mountain after the war and I got into the logging industry. I started working first around Shaver Lake and Blue Canyon. It was pretty hard work. I was a logger until 1983, when I went to work for State Parks and Recreation. First I went to Lake Tahoe; I worked there for about three years. Then a job opened up at Millerton so I transferred back home. I was a maintenance worker in charge of the Madera side of the lake—all the maintenance work over there. I liked it; it was a nice job for me. I saw a lot of river life during that time: coyotes, eagles, mountain lions, bobcats, and rattlesnakes. There were river otters too. People would sometimes go and shoot around the lake. They'd even shoot the otters. People will shoot at whatever moves.

I did all the railroad tie work that's down around the lake. I remember one night I dreamt about what I was going to do the next day. You know how sometimes you dream about your work? Well, I saw myself falling and I had railroad ties in my arms. I fell on the banks and broke both my legs—all this happened in my dream. The next day I was doing the work I had dreamed about, and I was pulling this railroad tie backwards. I remembered my dream and thought, "That might happen.

I better not be doing this." So I did the job in another way, to make sure I wouldn't have to live that dream. I've always thought about that, how that dream came to me, warning me of something that was going to happen.

I think that messages like that do come to us. My sister lived out by the river where the road goes back to where we used to fish. When they had that big earthquake in Tehachapi, her old dog and her horse went just about crazy before it hit. The horse was running around and the dog was jumping and barking; then all of a sudden it got real quiet and that's when my sister felt it rumbling, heard the noise. Those animals knew what was coming.

There's a strange story about how the bingo hall happened. My brother and I were the first ones to start the bingo hall in about 1985. We were having a three-day church meeting, and a Pentecostal lady from Fresno came up to the meeting. Nobody knew this lady at all but she got up and gave a message in tongues and it was interpreted. She said that there were people coming here from the east, the west, the north, and the south. That was the end of her message and nobody knew what it meant. Two months later another lady came up and gave us a message. She said that the Table Mountain area would be known as the *lighthouse*. She said that she saw a bear standing up on top of Pincushion Mountain and he had a barrel of honey under each arm and was running down to the people. Our preacher said, "This is something to consider; we should pray seriously because none of us knows what this is about, only God knows." We started having more meetings; there was another message, and another and another.

It was around this same time when a guy from Fresno came out and said, "I want to build a bingo hall for the people here." He bought fifteen acres up on the hill. He moved all his equipment in to level the place off, but he wasn't in our tribal trust. We told him, "Don't do anything yet or the county will come up here and they'll shut you down." That's exactly what happened. The county came in, red-tagged everything, and they stopped the project right then. My brother Ray said, "I've got some property down by the road; I'll give that to the tribe." And Lester Burrough had about three and a half acres also and he said, "I'll give that to the tribe, too." All of that property was in the tribal trust, so that's where the bingo hall went in.

We didn't have anything, no money at all, so we told the guy who wanted to build the bingo hall—we called him "the money man"—to build the hall on our property, right next to the main road where people could see it when they drove by. That's how we got the whole thing started. On December 27th of 1987 we opened for business.

We've had our up and downs—tribal members don't agree here and there—like all businesses. But it's survived. Bingo—that's how it all got started. It's been good for us and for Table Mountain. Another stage.

Lorraine Person

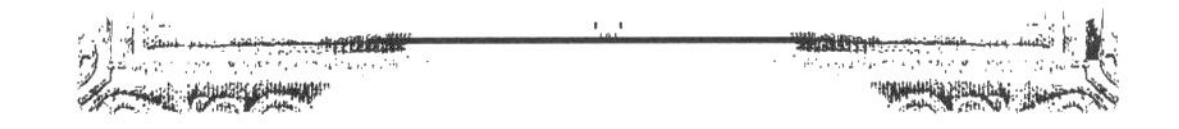

Lorraine Person sat on a folding chair one spring morning in the middle of a field near river property once owned by her great-grandparents and told of their gold-mining successes and failures, the scandals and romances, and of her own wide range of country experiences, from childhood to her eightieth year on the river.

MY GREAT-GRANDPARENTS, George and Joanna Wagner, were some of the earliest settlers in the Millerton area. I don't know all the details of how they came to be here, but I know that Great-grandpa was German-Canadian and he came across as a bushwhacker on the trains going west on the Oregon Trail and into California. Great-grandma came as a nanny to San Francisco. They met in the city but they moved up to Cascadel, north of here. He was a gold miner but she was afraid of the Indians. He finally brought her down here and they settled in Millerton. My grandma was the eldest of their five children, and she married a man named Henry Oiler.

I'll probably get into some hot water for telling this story, but I'm going to tell it anyway. Grandpa Oiler was a miner and he'd been up in Mariposa looking for gold. This would have been in the late 1890s. Apparently Grandpa hit the biggest gold strike that anybody had ever found up there. Many individual miners were all working in the same big mine, but if they found anything, they owed most of it to a man called Putnam, for store expenses and mining equipment. By the time they got through, Putnam got everything those poor miners had. So I guess Grandpa found the biggest vein ever found in the mine but he had also been working his own little mine just a short distance away. So rather than telling Putnam about his big strike, Grandpa just salted his own little mine with the gold that he'd found in the big mine. I hear Grandpa got about thirty-five thousand dollars out of that mine in Mariposa, but when he was found out, he quit mining real fast. Now, you can see that this is not a story that you want your family to talk about, but since Grandpa's long gone, I think that it's okay to tell it now.

Grandma and Grandpa first settled in Fresno. I never saw my grandparents dressed in anything but old common clothes, their country clothes, but I have a photograph of a very elegant lady and I had to ask, "Who is this person?" It was my grandmother. Their city life only lasted for a short time. My mother was born in Fresno; my mother, Agnes, was the eldest of their

"One day my brother and I had been swimming back and forth across the river, and as we were walking along the river, we saw a group of people getting ready for a baptismal ceremony."—Lorraine Person. Baptism of Ted Plett, early 1940s. Courtesy of Saundra Plett

five children. My grandparents homesteaded along the river, right in the middle of where Millerton Lake is now.

Grandma went into the cattle business with her brothers in the Wagner family, and as far as I know, Grandpa never did another hard day's work in his life after they moved up to Friant. I guess the cattle had hoof-and-mouth disease once, and Grandpa was hired to guard the bridge on the Madera side; he had to make sure people didn't go by until they walked through the sheep dip to prevent the spread of the disease. He was also hired as a guard for a while when they were building the dam; otherwise he just hunted and fished.

One year, right beside the bridge, Grandpa caught forty-five salmon. Forty-five! He became very popular with the people that came up here from Fresno because they always knew that Henry would get them a salmon. In fact, I came up one time and there was a lot of excitement. This fellow said to me, "You aren't going to believe this."

I said, "What? What?"

He said, "Well, Henry really caught the big one today."

It seems that Grandpa had thrown his fishing line up on the hill to dry and he had caught a rattlesnake with it. The dogs started barking and there was a lot of commotion. Believe me, that was the fish story for the day!

Grandpa hunted deer too and he never wasted one bit of it. He would cut all the meat up and he'd make jerky out of whatever they didn't eat right away. We used to take big flour sacks of jerky home with us—the real stuff; it tasted so good! He also trapped beavers and bobcats along the river; I remember seeing those and skunks too. I have his records with details of how much each of those pelts was worth.

My mother and her brothers and sisters went to Marysville School. It's gone now but it was there for years and I think the old trees are still there. Later they all went to Clovis High. My mother drove the car to school. My mother was quite the driver. She drove Ed Brown to high school as a freshman. He lived along the river and she'd pick him up in a Model T Ford. Ed always said, "Lorraine, the wildest rides I ever had was when I rode with your mother to school." The other kids took over driving after Mom dropped out. She only went to one year of high school.

My mom was only seventeen when she met my father. He was a Greek immigrant; his real name was really Partheneus Andronico, but he was given an American name, Pete Andrew. When he first came to America, he was working in Fresno raising grapes. You can actually still see remnants of the vineyard that they had along here. The story goes that my father swam across the river to meet my mother—that's what they always said. My mother was camped on the other side of the river and my father swam across. My father talked my mother into eloping with him when she was seventeen years old. My father was ostracized because of that elopement. He and Grandpa never did talk, for as long as they lived. My mother and I would come up to see Grandpa Oiler all the time, but never with my dad. We lived in Madera County, in the Ripperdan district; that's where I grew up. My fondest memories, however, are on this river and those are the stories I want to tell you about.

We came up to visit Grandpa frequently. Mother would bring him fruit from the valley and other stuff he needed up here. Getting up to Grandpa's house wasn't the easiest thing, especially after a flood. My grandparents had a creek on their property that fed down into the river. My mother would grit her teeth—I can still see her now—and she would back that car up, and we'd dive into that water and sashay across. I was holding on for my life. I just knew we were going to drown or be washed away.

I loved to go with Grandpa on the trail that led from his house, in the middle of where the lake is now, down into Millerton. There were some Indian burial grounds and we'd find arrowheads and little glass beads. The only Indian family that I knew much about was the Hudsons. Daisy Hudson was the Indian girl that my mother played with. Mom and Daisy were good friends. I remember a story about one day when Mom was out riding horseback, herding cattle, and Daisy invited her to come over for lunch. Daisy's house was along the river; I remember their little village there. When Mom got to Daisy's her mother was pounding something and Mom asked, "What is that, Daisy?" Daisy said, "That's grasshopper. Grasshopper legs are very good." Mom said, "Oh, you know, I just remembered that I didn't tell my mother where I was and she'll be very worried about me." She jumped on her horse and rode home. My mom didn't want to eat those grasshopper legs! Uncle Eddie was friends with Daisy's twin brothers. He said that they were the ones that taught him how to use the bow and arrow but they used to fight with him a lot too. He said they couldn't beat up his older brother, because he was quite a bit bigger than the twins, but they'd beat up on Eddie, even though they were all good friends. You know how boys are.

There were rattlesnakes from one end of the place to the other. Every time you came up to the place, you ran into a rattlesnake. Grandpa always had a couple of dogs and every once in a while the dog's face would get bitten, and he would run down to the river and rub in the mud for a day or two. Grandpa sold some of the rattlesnake skins. He had some great big ones. He'd put them on a board, let them dry, and sell them.

Grandpa made his life on the river. He mined for gold too, all up and down this river; there was still a lot of gold on the north side. I'm not sure to whom or where he sold that gold, or for how much, but I still have his gold scales. He did a real business with it.

There were benches and little cottages across the river and we'd see people picnicking down there quite often. One day my brother and I had been swimming back and forth across the river, and as we were walking along the river we saw a group of people getting ready for a baptismal ceremony. We stood on the hillside looking down at them and they said, "Why don't you come and join us?" We said, "No, thank you. Thank you, anyway." Pretty soon they started baptizing people, and it seemed to me that they practically drowned them. I kept thinking, "Bring them up, bring them up! How long does it take to get the spirit?" When they got through with the baptizing they pulled out all their big larders of food. They went over to their little old rattletrap cars, brought food out, and put it on the tables. As we stood there, they said, "Would you like to join us?" We said, "Oh, yes, thank you." You see, we said yes to the food but not to the dunking. We ran down the hill, jumped into the river, swam across, and had one of the best feeds ever. They knew how to put on a feast!

When the salmon came up to spawn, we would come for picnics and then we would swim with the fish. They were just swarming around, huge groups of them. We would get into the river and, because they were kind of lethargic, we could get hold of their tails and let them pull us along. We'd try to get *on* them and ride them, but of course that didn't work out too well. I guess we didn't have any fear at all because all I can remember is the fun of swimming with the fish. At the time I didn't realize what a rare thing that was. Now I look back on it and say, "I tried to ride on the back of a huge fish. I swam with the salmon."

I was volunteering as a docent for the Parkway a few years

ago when we had a big flood. The river kept getting higher and higher; it came all the way up to the restrooms at Lost Lake Park. One little boy came dressed up like Daniel Boone that day. I said, "Oh good. We're all going to be Daniel Boone today because we don't have a trail anymore. Our trail is in the middle of the river." As we were watching the river, the merganser ducks came flying by and hit right where the middle of the river should have been. The water just whooshed them downstream. They'd come back and hit the middle, over and over. The water was so fast and so high, there was no place for them to settle. I felt like Daniel Boone too. I mean, how lucky can you get? I feel fortunate, at my age, to be able to do things like that—to be able to come right back to this same river and share it with the generations to come.

My heart is tied to this river. The dam changed it all, it really did, but the main thing that started to bother me was watching all those new houses coming in—encroaching, encroaching. I went down to city hall for one of the first hearings about the Ball Ranch development and I talked about why I didn't want houses there. I told them my story. I have a lot of memories on this river—from the time I was a little girl to much later as a docent. All the money in the world couldn't pay for the basic things that this river provides. I want this river for my grandchildren, and my great-grandchildren, and my great-great-grandchildren. That's my guiding premise.

Fred Biglione

Sitting on the south bank of the San Joaquin River, directly across from the now infamous old broken Friant Bridge, Fred Biglione, a retired general contractor, told his story of roaming the riverside country with his brothers: the floods, the fishing and hunting, the hard work, and the fun of it all.

WE WERE UP HERE QUITE A BIT the year that the Friant Bridge went out; it was exciting to watch the river during the floods. The water was to the top of the bridge; big trees were coming down and they would mash the pipe railing. We watched the flood while dodging trees. Amazing. Then the next thing you know, the bridge was down. We weren't there when it actually fell—when it washed out—but we saw it right after it happened. A lot of water came down that year. This was before Millerton Dam was built; it was in the late thirties.

We had a home on Copper Avenue and Auberry Road, so we were close enough to come to the river all the time. We used to roam for hours and miles. There were no houses. It was just us, roaming the country, three boys hunting and fishing. We were careful about never shooting anything we didn't eat; that's still a family rule. I've hunted all my life but I never shot anything that wasn't in season or edible, whether it was a raccoon or a skunk or a coyote. Of course, I have coyotes on my property now, but I wouldn't think of shooting them. I love to hear them at night; it's such a beautiful sound. We never ever hunted any deer. There are plenty along here, though. Of course, the Ball Ranch was always closed to any shooting at all, and we respected that.

Later, when my sons were growing up, their big challenge was to go down into Little Dry Creek and get onto the Ball Ranch. They wanted to catch some fish they had seen, some big trout that they spotted in one hole. One day, Willis Ball saw that someone was out there, and my kids saw him. So they jumped in the river and got under a cliff, and they heard Willis saying, "Well, I know I saw somebody," and he kept looking and looking. They were right under him. Later I said, "Boys, you're going to get caught up there. You better be careful; the Balls have eyes for trespassers." The Ball Ranch was always taboo; you just didn't go in there. Everybody knew that.

To get legal access to the river when I was a kid, we had to go down to Lane's Bridge at the end of Copper Avenue. There were a lot of people that went there for wiener roasts; everybody did that. There were also big sand beaches for swimming. Lane's Bridge was the closest beach to us, and

Two youths and their dog on the Minarets and Western Railway, where Friant Dam now stands. Courtesy of Eastern Fresno County Historical Society

that's where we went most often. We canoed the river long before other people were doing much of that. We didn't have money for a powerboat but we had five canoes, which were more fun than a powerboat anyway.

We caught salmon down at Lane's Bridge. We didn't have any money, so we had the simplest pole in the world. Just worms. The males would be the only ones that would actually strike, because they wanted to fight off anything that was interfering with the females laying their eggs. The male salmon were attracted to anything that seemed to be an obstacle to spawning, so they fought off our bait, and that's how we caught them. We never caught any fish we wouldn't be eating, though. We always ate them; we marinated them first, then we smoked them with salt and molasses.

I remember the train coming up here to Friant and the wooden trestle that went across where the dam is now. In fact, we rode across that thing on a one-horse cart we had. Our mare was so good that she could actually walk on those railroad ties. She was one of my granddad's work mares but she was more slender, like a saddle horse. She was our best; she took us everywhere pulling that two-wheeled cart. It was like a sulky, and when we were young, there was room for three of us on the seat.

There were just a few families who lived along the river and we knew them all—the Ranks, and the Wagners, and Tony Stoten. There was the Japanese family, the Nishis, and a few other Japanese families below them. The Nishis lived where the vineyards are now. Bill Rohde took over their property during the war when they put the Japanese into internment camps. Certain people in Clovis took over all the Japanese ranches, and not all of them got their land back after the war. Some of the people who took over kept the land, claiming that they had farmed the property for so long they had the right to keep it. To this day, I've lost all respect for those people.

My grandparents, all four of them, immigrated from Italy. My mother's parents went to Northern California in 1895 and my dad's parents came to Clovis in that same year. They didn't have anything at all. My mother's parents came from the Alps of northern Italy and they wanted to be up in the mountains, so they ended up at June Lake. My granddad worked on a mill; in the pond, pushing logs to the saw. When it was time for my mother to go to school, they moved down to Clovis. And lo and behold...my father's family lived on Fowler just south of Shaw.

Being Italian wasn't good in those days; they had to keep my father home from school because of the prejudice he faced, so he never finished grammar school. But my mom did; she graduated from Clovis High School. My mom was born in 1899 and my dad was born in 1900. My dad was born on October 18th and my mom was born on October 19th, so they'd be the same age for one day, and then the next day she was a year ahead of him again. We kids liked the idea of that.

I remember lots of people would stop by my granddad's because he made wine. We kids got to meet and talk to them. There were two fellows I'll never forget. They were brothers from Spain and they lived way back on the road behind Marshall Station. They came to town once a week. They would get their wine from Granddad and joke and whatnot. I really admired them because they were different, interesting. There was another fellow we called Happy Cat. He lived up at the base of Mount Owens and he used to go to Clovis by horse and buggy. He would stop by my granddad's, too, on his way back from town. He'd get his wine from my granddad and wind up spending the night. He really was one happy cat.

My parents had a small acreage of fruit—grapes, apricots, and figs—but mainly we dry-farmed grain. We didn't have anything; we just leased the land. Later we had cattle, but not until I was about twenty; we just dry-farmed before then. This was around the time Wiley Giffen lost everything across the road from us. The family had to give up their land, and the bank wanted my dad to take it, but he couldn't. My dad couldn't even take it for free, because he couldn't pay the taxes. He was afraid to take the chance.

I was always proud of how hard my family worked. I liked hard work. As a farm boy, I was always interested in building things, even when I was just a little kid. I watched the building of the dam with fascination. I was a teenager when they started it. They brought crane loads of equipment in, and their method of unloading was to get up top and drag the equipment off. If you could do that, you had a job. I remember some older guys from Clovis that I was real proud of; Cliff Wilson was one of them. He got up there, started the D6 engine on the train, and backed it right up. The foreman asked, "You have a job?" Cliff said no. The foreman said, "Well, you got one now." They hired him on the spot. I was so impressed.

My brothers and I grew up working hard. We enjoyed every minute of it; you couldn't stop us. We worked on my dad's twenty acres and then we worked on the neighbor's place. We were eight, nine, ten years old, picking peaches. The trees were so much bigger in those days than they are now because farmers pruned them differently. We'd climb up fourteen-foot ladders, working right out there with the adults. We made twenty cents an hour. We saved every dime, literally every dime. When I got married, I had enough money to build our house. I built it myself, of course. I installed the flooring, painted it, and furnished it with money that I made at two bits an hour.

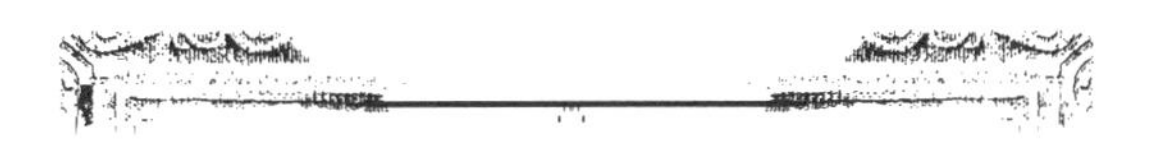

Brooke Wissler

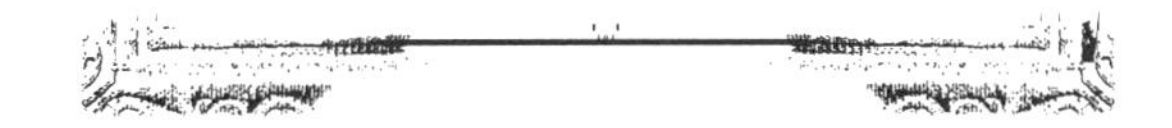

In her lovely old farmhouse surrounded by a forest of elm trees, Brooke Wissler took us on a tour of her family photo album, an amazing visual history of her valley ancestors and the river they settled near, the San Joaquin. She is an active volunteer for several Fresno art museums. Her late husband, John, was a founding board member of the San Joaquin River Parkway and Conservation Trust, and now her daughter, Cecilia, serves on that same board.

MY GRANDFATHER CAME OUT HERE in his late teens after the Civil War. He was in the Confederate army. After the war was over, the period of reconstruction was hard on the South; it was a terrible time. First, my grandfather went down to work for an uncle who had a farm in North Carolina. When that didn't work out, he got on a ship to the Isthmus of Panama and went by mule across the isthmus. Then he went up the coast until he landed in San Francisco; from there he went to Stockton and then down into the valley, where he'd heard there was cheap, cheap land. He started out living in something called the Alabama Colony, which was up to the north and slightly west of us. It was peopled by southerners who had just come out west. I think my grandfather had no conception of what it was really going to be like, because this was a pretty barren land. But he stuck it out and he married my grandmother, who was also from the South, from Mississippi. Her father brought the family west after the trains went in, and they also ended up in Fresno. My grandparents met and were married and, as a wedding present, they were given this ranch for six gold pieces. This would have been in 1870.

My grandparents had sheep and cattle and they did some dry-farming. They had five children, my father being one of them. My father came back after the First World War and took over the management of the ranch. They still had sheep

Posing on a handmade raft: Brooke Wissler's father, Brook Mordecai, top left, and Brooke's mother, Caroline Preston, bottom left; circa 1910–1920. Courtesy of Brooke Wissler

Wading under Skaggs Bridge. Brooke Wissler's aunts Ethyl Fleda and Louise Mordecai with a family friend, circa 1910–1920. Courtesy of Brooke Wissler

and cattle, but once they built the dam and got irrigation into this valley, he switched totally from dry-farming to alfalfa and robust cotton.

I'd ride my horse across the land between here and the river, just green open land all the way from here to there, and then I'd ride down into the river. There was a lot of water and wonderful vegetation along the banks. The sycamore trees...oh, they were beautiful! My uncle and his wife built a house up on the bluff and they owned one of the islands on the river. They did some clearing on the island and we'd camp down there; it was lovely. This would have been in the mid-to-late thirties, way before the dam, and there was still wonderful fishing. We didn't fish but lots of people did, and there were quite a number of men who worked on our ranch that would go over and catch huge salmon. The only problem that I remember on the river was the stinging nettles. We used to have to change our clothes behind the bushes, so we had to be very careful about those nettles. That's the sort of thing that kids remember, but really, the place was absolutely enchanted.

The whole area was so scenic back then, so spacious. In fact,

Maynard Dixon, the famous California artist, was my father's cousin. He was a very prolific painter and he used to spend quite a bit of time out here; he loved it. We still have some paintings that he did on the ranch. He was married to Dorothea Lange, the photographer, and she did some wonderful photos here at the ranch of us as children. My father and Maynard were first cousins and they were close, but Maynard and his wife were *very* San Francisco, very bohemian. My father, bless his heart, he was a little square—although he *was* a Democrat. Maynard would come down with Dorothea, and they'd go over to the cattle camp and get very chummy with each other. My father thought it was inappropriate, but I thought it was wonderful. Really, what a childhood I had!

I love living on this ranch. We're in a perfect location, with Cottonwood Creek running through and the river so close. My husband, John, loved it out here too. He was a great outdoorsman. He was very conservative politically, and he and I used to have terrible arguments about that sort of thing, but as far as preserving the river was concerned, he was devoted to the cause. He believed that people should have better access to canoeing and be able to hike along the river. I'm sure he would have approved of sending more water down the river. That would have been very exciting to John; I wish he could have seen that. It looks like that may really happen now. Won't that be something?

My daughter lives out here too, with her husband and children. We've managed to keep this property in the family for a long time. We have a lot of family history on the river. You can really feel the history when you see our photographs. I have a couple of old pictures of my mother and my father when they were courting on the San Joaquin River. I have a photo of my mother and my two aunts spending the afternoon at Skaggs Bridge. I have one shot from 1927 that shows our horses in the river, and a picture of us on a raft in funny bathing suits. My mother and father look like they have their underwear on! Oh, and look at the white dresses. Amazing that we have these photos, isn't it? Lots of history.

Louise Mordecai (Brooke Wissler's aunt), circa 1910–1920. Courtesy of Brooke Wissler

CLAUDE P. DORMAN

Elmer Hansen

Water was always one of Elmer Hansen's primary interests in life, from his love of rivers to his long career as a founder of Peerless Pumps, one of the first companies in the San Joaquin Valley to pump water from the ground and forever change the way we live and grow our food. He was ninety years old when he told us about his life and the monumental changes he had witnessed. He died soon after, leaving us with his fascinating story.

I REMEMBER BACK WHEN I WAS a little kid there was no electricity, and everybody had to hand-pump the water out of a well. We carried the water in buckets to the house to wash the dishes and heated it on the stove to wash our clothes. The washtub was our bath; we stood up in it, and mother sprinkled us with a wet rag and washed us from the head down. You had to carry water to the cattle; you had to carry it into the house. If you had a family vegetable garden or a rose garden—all that would be hand-bucketed.

I moved up to the San Joaquin River in 1929 before we had the dam at Millerton. The river looked entirely different in those days. I've spent a lot of time on rivers, pretty much my whole life. I grew up in Reedley, swam at Reedley Beach, and at a great swimming hole on the Kings River, just above Piedra; it's now called Winton's Cove. I was always interested in fishing and boating and swimming—and girls, of course. A river is a great place for all of that.

We bought the ranch where we are now in 1957. We had roughly a hundred brood cows, and we raised calves, but that market really took a beating and we got out of the cattle business. We've had walnuts since 1971. We left the land natural, never leveled it, and it's all under drip or fan irrigation. Originally there were five pumps on the property for sprinklers, but now we're down to three pumps under the drip arrangement. With the huge power bills we have now, that was a wise move.

The irrigation districts in the valley were formed to help supply water for the grape vineyards that the Italians started planting here in the valley. The more vineyards they planted, the more water we needed, and that's when I came into the picture. I became involved with Peerless Pumps in the early thirties when we started installing deep-well pumps.

By that time, irrigation had already lowered the normal water level by three to five feet. The valley is about 325 feet above sea level, and the gravity lift is right around twenty-eight feet. That's the point where you lose prime on the pumps. So we had to start deepening the wells and get the pumping pillars to *lift* the water

"I would say that in my forty-seven years with Peerless Pumps we installed over forty-thousand pumps in the valley."—Elmer Hansen. Image from the Peerless Pump collection, courtesy of Peerless Pump

instead of *pulling* the water, to overcome gravity. It was new technology.

Pumps changed everything for agriculture; they made it possible for this county to be what it is. First, new grape varieties were introduced, then citrus and nut trees, and they all had to have water. Even after the dam was installed, you weren't assured enough water to irrigate through the entire growing season, not in the summertime. So farmers had to supplement with these well pumps to finish out their crops. I would say that in my forty-seven years with Peerless Pumps we installed over forty thousand pumps in the valley.

I've been retired for years now, but I've kept my interest in water, and especially in this river. I hope that somebody is going to carefully watch what's happening. We have to make a strong plea to keep at least one hundred second-feet of water running in the river year-round, because when you cut it back to fifty or below you're in trouble. A lot of the old trees, the sycamore and the oak, are dying. The river itself is filling up with willows and bamboo. I don't want to point fingers, but I am worried about the wildlife, too. Our whole neighborhood has dogs and cats that are destroying all the cottontails, and I think we're going to have a real problem with some of the larger birds, like egrets and the blue herons. They don't like people. The ones that were nesting in the sycamore trees along our property have moved somewhere else; they're not there this year.

Thankfully, we're still on a flyway between Millerton and Fresno. We had a mallard duck and his companion in our swimming pool this spring, looking for a place to nest, but we discouraged that move; with the chlorine in the water we thought they best fly on. I counted the other day and we have twenty-two Canadian honkers that fly right over our place. At night they nest in one of the San Joaquin lakes, but they leave early in the morning, before people sic their dogs on them because they honk all night. That's beautiful for me to see, and to hear. I think a lot of birds are getting their food at the Fresno sewer farm, especially the sandhill cranes. They come by the thousands every evening in the fall and spring. Thousands of them go north every evening from about four o'clock to sundown. It's just beautiful to watch.

We've lost the badgers and the beavers. We had a fire through here five or six years ago, and the cottonwood trees that the beavers like to build their dams with were all destroyed, so the beavers have moved on. We occasionally see deer. I guess the deer are pretty rough on golf greens, with their sharp hoofs. It's hard on the animals, all the activity, all the people moving in on them.

There weren't many people out here by the river until pretty recently. I'd say I used to know 80 percent of my neighbors, all up and down the river. Now we don't know most of our neighbors. A lot of the old-timers have either passed away or their kids weren't interested in farming, so the land has been sold off to people who know nothing about this area, nothing about farming or the river.

Our farm will be a big beautiful park one of these days. It's under contract to the American Farmland Trust and can never be developed into subdivisions. I wanted to be sure that I locked that in. I imagine it might have opened a door for other people along the river here to consider doing the same. There are lots of people interested in similar arrangements now that we can all see the writing on the wall.

I've got a lot of family history here in the valley. In fact, there's a big picture at the Bank of America in Clovis of a harvester in a grain field—that's my dad out on the deck. He was known as a jack line driver, with a lead horse and thirty-six mules pulling the harvester. That was taken on the ranch

"Pumps changed everything for agriculture; they made it possible for this county to be what it is." —*Elmer Hansen. Image from the Peerless Pump collection, courtesy of Peerless Pump*

that was deeded to my dad and uncle by their great-uncle. We go way back. I've seen a lot of change. I've been through two world wars; I was born when there were no telephones and no electricity. My first radio was a crystal set run with a dry-cell battery. Even yesterday, I was visiting with some of the old-timers and we were talking about some of the changes we've lived through. There was the Victrola, then black-and-white TV, then color. Now there are laptop computers. At the banks, you're just known by a Social Security number. It's a complete change. I'm now ninety years old and the doctor says I've got another ten to come, so maybe I'll see even more changes. Who knows?

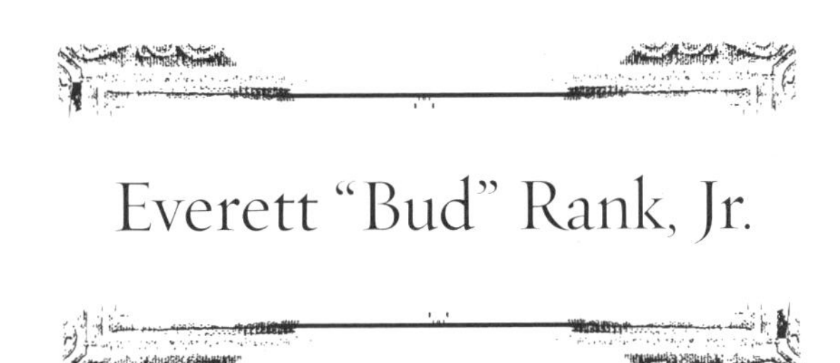

Everett "Bud" Rank, Jr.

Any collection of stories about the San Joaquin River must include Bud Rank, who closes the circle of history on this upper stretch of the river. For eighty-four years he has lived below the bluffs along the river. His story includes not only his childhood, before the dam was built, but also the important role his family played in the epic legal battles for riparian rights fought by his father and the local farmers for decades.

MY GRANDFATHER BUILT HIS HOME on the river. It was not majestic at all, or even a particularly nice home, but he had a pretty white border fence around the whole place at that time. It was surrounded by vineyard, all irrigated on the ground. There weren't many farms on the river in those days, but he had a good one.

I've lived in this river bottom for about eighty years. I was born in a house just a half-mile from where I still live, on the same property, in the same room where my wife and I slept after we got married. I've got a lot of memories here.

During the Depression, when I was little, we had to raise everything we ate. We had chickens and eggs and a cow to milk. We would go fishing on the island, just to get something to eat. I remember I thought my mother was mean for making us do that, sending us down to catch a bluegill for dinner and making us pick berries, a lot of berries, so she could can them. It felt like work to me then; we did it to get something to eat rather than just going out to have a good time. It wasn't for fun.

My dad loved to fish. He'd spend hours and days down on the river. There were king salmon—the largest species—that came up here; they were huge. My dad and his friends would go down the river in a little boat with a motor on it. Once he caught a fifty-pound salmon, even though they were only allowed to keep up to thirty pounds. He would can the leftover fish. They would eat all of it, the whole salmon.

When I was a really little boy, maybe six or seven years old, all of my dad's friends fished with spears at nighttime. Many, many nights they did this. They'd put a tin reflector on the side of the boat and an old Coleman lantern under that, and the reflection would shine into the water. They'd go sideways down the river with spears. Spearing was strictly illegal but my job was to go down the river, stay ahead of the boat, and watch for the game warden. And we never got caught! I don't really know why the spearing was illegal. I guess it's sort of like the laws that say you can shoot birds, but you can't trap them. With the spearing, the fish didn't have a chance. With a hook, there's more of a game to it.

The Indians used to camp here during the salmon runs. In

Everett "Bud" Rank, Jr., age seventeen, on his horse Chief. Courtesy of Evelyn Rank

those days they'd walk down from Table Mountain with a horse or two to carry their things. I had no friends around to play with when I was a little kid; there weren't many little children down here. I'd go down to the river and listen to one of the old, old Indians; George River was his name. He must have been eighty or ninety. He'd tell me about his background and the life of the Indians. It was really interesting to hear him speak, and I got to watch how they lived. The Indians would dig reeds to make baskets. Rank Island was one of the good places to get reeds; there's an area there where they really grow thick. The Indians would dig those and bundle them all up. I watched them make baskets, too, many times. There would be at least fifteen men and fifteen women and many children camping by the river. They would come down in the spring through the early summer to catch salmon; then they would come back in the fall to pick grapes for us. They were great, great pickers and the most honest employees we ever had. We paid by the tray and they would completely fill those trays up. They were very dedicated people. I got to know all of them who lived up in the mission where the gambling place is now. I still know a lot of the Indians, the ones that are still around.

It was the mid-1930s when the Bureau of Reclamation started coming around and conducting surveys, and we got an indication of what was coming. I say *we*, I mean the farmers and ranchers along the river. I mean my dad, my family. We started to understand what they were planning on doing with the dam. It disturbed the farmers because they believed that this was their property, that they had a right to pump water out of the river, and to fish, and to keep the river as it was, a beautiful and majestic stream. Back when this area was first homesteaded, the old laws said that land extended to the center of the river. That's how the original title shows ownership—all the way to the center of the stream. But then the government was planning on damming it all off, stopping *all* the water down the river. That was a terrible thought for the farmers—no river water at all.

I think it was in 1938 when all the neighbors gathered at the old Fort Washington School, and that night they voted to sue the federal government. Now, money was tight then; the Great Depression was still going on, but they did collect some money and my dad was elected president of the group. They named the suit *Rank v. Krug*, although other neighbors were involved too. There were about five neighbors who committed themselves, who said they would see it through. Later on, by begging and hoping, they got other landowners all the way down to Gravelly Ford to sign up. They had a lot of people involved but not much money. As the suit went on, it cost them more than they ever thought it would.

Everett "Bud" Rank, Jr. amid his cotton. Courtesy of Evelyn Rank

The farmers' lawyer had worked for the city of Fresno; he'd been the city attorney and he had friends on the city council. He advised them to join the suit, so the city joined the *Rank v. Krug* suit and that put another lawyer in court. Eventually, that's how Fresno got its water. The International Irrigation District also joined the suit—Danny Andrews and his group of farmers. They went to the federal irrigation district and asked the district to join the suit too, to put up some money, but they said, "We don't need no more water. We got all we want; we don't want to join your suit." That was a major mistake on their part; they let all their water go down to the southern part of the valley. But I can see how that happened; it was hard times for all farmers then—trying to fight the federal government and at the same time feed their families. They were courageous people, my dad and the neighbors here.

At first, the state of California was on the farmers' side.

They had a lawyer in court fighting for water rights and the maintenance of a stream for the salmon. But there was an election for governor in the middle of the whole lawsuit. The day after the election, the state's lawyer moved from the farmers' side to the government's side. They reversed their position and said, "We don't need water for the fish; go ahead and let it dry up." The farmers were fighting the case hard, and losing the state's support was a big blow. The politics were frustrating, but they kept trying. They went through all the courts, and in every court they went through, they won.

Finally the federal government repealed all of the rulings and the Supreme Court threw the whole case—*Rank v. Krug*—out because of a technicality. The Court ruled that the farmers hadn't gotten the proper permission to sue the federal government. When it all came down, the ruling wasn't on whether the farmers were right or wrong, but on a technicality. I'll never forget that because the Supreme Court justice was Earl Warren, our former governor. He was back there in Washington and he ruled against us. It was a sad day for the farmers of the San Joaquin River, after fighting for all those years.

In the end, the Bureau of Reclamation did guarantee the farmers some water. The bureau promised us water at the southernmost part of our farm—five second-feet—which was enough so that we did have some water for our pumps, and there was still some stream running down the river, all the way to Gravelly Ford. But they almost dried the river up many, many times, just to send the water south.

The first federal judge hearing the case, Judge Pearson Hall, was really a great, great person. He believed in the farmers' rights; he didn't believe that the federal government should come and take over the water and destroy the stream. I remember one time he held the trial down at the riverbank, right near where the canoes put in. He was trying to show what this river was going to look like when they dammed it. He said, "These little trees are all going to die; you won't see any more sandbars. When you put the dam in, it's not going to allow sand to go down the river from the mountains. The dam will stop all that." He was right. If he was alive today, he'd say, "I told you so!"

The river looks okay now, but not like it did, not at all. It's a lot deeper now. By keeping the rocks and silt from coming down the river, it all stays in the lake. So the river bottom is eroded out; in a lot of places it's about five or six feet to the bottom. Our groundwater levels here have dropped by fifteen feet. And, of course, the salmon have disappeared.

The sad thing was that the farmers had a solution that would only have cost about $2 million. We thought it was a great compromise. We had hired an engineer from San Francisco, a Mr. Lee, and he came up with a great cheap plan. He proposed a series of check-dams in the river, at different elevations. Those check-dams would be up only during the times when there was a low flow, and they would back the water up so the trees and the shrubs would always have plenty of water along the banks. When the high water came, the engineers would lay those dams down and the water would flow all the way to the ocean, free. Today we would have a river of which we could really be proud. Instead, there are a lot of areas that have completely deteriorated.

The last salmon run we had was in 1947. My good friends Bill Rohde and Harvey Moore came down one night. The three of us put the boat up the river and we speared salmon; we filled the boat up—honest. Oh, what a night we had! That was the last time that salmon ever came up the river. It was dry then from Mendota, from Gravelly Ford on down. They put the dam in and that was it.

"Then I thought, *What the heck are you doing?* I mean, that poor thing. This is his home. After that day, I wouldn't shoot anything." *—Everett "Bud" Rank, Jr. Courtesy of Evelyn Rank*

Everett "Bud" Rank, Jr. (pointing) and his father on the island. Courtesy of Evelyn Rank

The Bureau of Reclamation was hard to work with. There was no environmental protection at all, nothing. Environmental impact reports didn't exist. I later went to work for the bureau; I worked up there for two years after I graduated from high school, on the survey crew, so I'm not blaming the people that worked for them. It was the people in Washington making the decisions. They sure couldn't get away with it today.

The length of all those lawsuits, from the time we started the suit to the end, was about fifteen years. I have the whole *Rank v. Krug* file from day one to the end of the trial. I love history. This area has all kinds of old history. This was one of the first Spanish land grants in California—this whole area, from Friant to Skaggs Bridge, from bluff to bluff. This was when Mexico owned California. The Mexican president gave this land to a fellow named General Castro for some great deed he did in Mexico. Apparently he had a fort he built down in this area and he raised horses. The name of his ranch in English means something like "The Ranch by the River San Joaquin." I don't know what happened to the poor fellow; I forget how they got the land grant away from him. I think the American government cheated him out of it in some way, just like they cheated the farmers when they dammed the river.

It's hard getting people interested in preserving the beauty of nature. Not long ago, I was over at Rank Island and there was a pickup over there. I knew these guys were up to no good, because there were some empty gun scabbards in the back of the pickup. I yelled and yelled and nobody answered, so I just went over and took all the air out of the four tires. I went back, called Fish and Game and said, "There are some poachers out here and I think they're hunting deer." Fish and Game came and arrested them; they'd already killed one doe.

I've got to say this. I used to be a deer hunter. I loved to shoot deer up in the mountains. One day I was over on Rank Island and I saw about four bucks jumping over the fence. I thought, "Man, I'm going to get that big buck." I ran home, got my rifle, and found one big buck on the upper end of the island. I got him up in my sight and I was ready; I was going to shoot him. Then I thought, *What the heck are you doing?* I mean, that poor thing. This is his home. After that day, I wouldn't shoot anything. It was the last time I went hunting. I didn't need to do that anymore.

Bill Wattenbarger

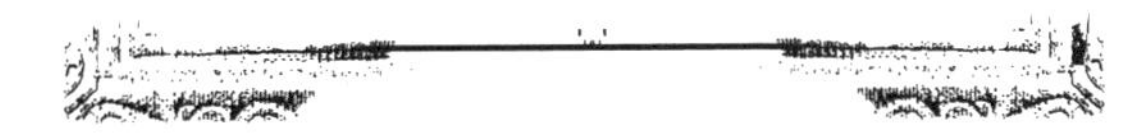

Bill Wattenbarger farmed almonds and raised his family by the San Joaquin River, on the same property that his father farmed before him. He still lives there with his wife, Anna, a longtime member and officer of the Parkway Trust's board of directors. Bill is an active conservationist and leads nature walks along the river and in the Sierra, with children and adults. His family farm is in a conservation easement held by the San Joaquin River Parkway and Conservation Trust.

My family came by train from Ephrata, Washington, to Madera in 1914. The Hayes brothers had already been on the San Joaquin River for two or three years before my folks moved here. My family bought some of their property and they were our closest neighbors. For years there was one piece of land between us and the river that still belonged to the Hayeses, but when they got old and decided to sell, we bought that portion of their ranch.

The last really major flood of the San Joaquin River was in 1936, the year I was born. Of course, there had been many floods before then, but after they built Friant Dam the floods were never that severe again. Many of the big water projects that made this area what it is today were started because of that particular flood. One guy who used to work for us, John Morrison, was a Dust Bowl refugee. He moved out from Oklahoma and the first job he got was building levees in Firebaugh, because in 1936 the whole west side was underwater. Even the Fig Garden area of Fresno was flooded that year; not by the San Joaquin, but by Big Dry Creek, Fancher Creek, and Red Bank Creek.

They started putting flood control on those creeks after that year. Local government stepped in and said, "We've got to end this problem." That's when they started the project to build Friant Dam. It was originally owned by the Madera Irrigation District, but they made a deal to give the dam site to the federal government. Madera didn't have the money to build the dam but the federal government did; this was during the Roosevelt era. So it was finally the Bureau of Reclamation that built the dam.

That's when the federal government started buying farmers' water rights. They didn't force anybody, and a few people held on, but nearly all of our neighbors sold their rights. For most people, it was just about the money; they were all willing sellers. They gave up their right to pump out of the river, but in place of river water, the government promised to build the Madera irrigation system. People thought, "We won't need our river water if we've got a canal."

When my parents first came, it was all dry-farming in the area. Washington had been dry too, so they did the same kind

Bill Wattenbarger's grandmother Ellen Wattenbarger with her nephew Virgil and her son Roy, circa 1918. Courtesy of Bill and Anna Wattenbarger

Built in 1935, the family home retains its place on the Wattenbarger Ranch today. Courtesy of Bill and Anna Wattenbarger

of farming here. They didn't have pumps in those days so mostly it was grain, barley and wheat. People also had horses, sheep, and cattle. It wasn't until the 1920s, with the invention of the deep-well pump, that irrigation was introduced. My parents had vineyards once they had irrigation.

I can remember some of the salmon fishing when I was a kid, before the water was cut off. I remember my dad telling one story about this guy going down the road on a motorcycle, and he had a salmon laying across the handlebars; it was the whole length of the handlebars. I remember the people fishing shoulder to shoulder off Skaggs Bridge, and when the water went down in the late fall we would see the dead salmon floating in the river. That was something to see.

I have very clear memories of later, in the early 1950s, when the eels came up to spawn just like the salmon. They would spawn in the gravel beds. This was before the water was completely shut off, before Friant-Kern Canal was built. I think it was probably in the spring; I remember walking out on those gravel bars with hip boots and all these eels scurrying between my feet. They were maybe a foot to a foot and a half long, and an inch in diameter at most. I think they were lamprey eels—freshwater eels, they call them. The gravel beds are gone now and so are the eels. The mining companies came in, so the gravel bars were completely destroyed.

Before the Delta-Mendota and Friant-Kern Canals were built, the river was still pretty wide. During the summertime, there was a full flow and there was a lot of powerboating. I remember one time a bunch of us little kids were swimming and we waved to this guy. He pulled over and gave us a ride. He had this little blue boat and he had on his bib overalls. All of the kids piled into his powerboat and roared up and down the river. We were so excited.

It's funny what you remember. One time a guy came around and wanted to trap some raccoons. He was an old guy and he knew how to preserve them, just like the old days. He'd skin them and put them on a stretcher. He gave us two raccoon skins for letting him trap on our property. That was a memorable moment. The ranch directly to the east of us was once owned by Jack Dempsey, the prizefighter. This was in the 1920s, after he'd made his money as a prizefighter and retired. He was kind of an absentee owner, but the ranch was known as the Dempsey Ranch for years after he sold it. That ranch has changed hands three or four times now; people are moving in and out. On the Fresno side of the river, there's a lot of development going on. But there are still people who have been here for a long time; some things stay the same. Of course, now there are no salmon or eels, but there are some big largemouth bass; the bluegills get quite large, too, and people are into the catfish. There are some carp that look like torpedoes down there. The river still has a lot to give.

"It wasn't until the 1920s, with the invention of the deep-well pump, that irrigation was introduced."—Bill Wattenbarger. Image from the Peerless Pump collection, courtesy of Peerless Pumps

BETTINA

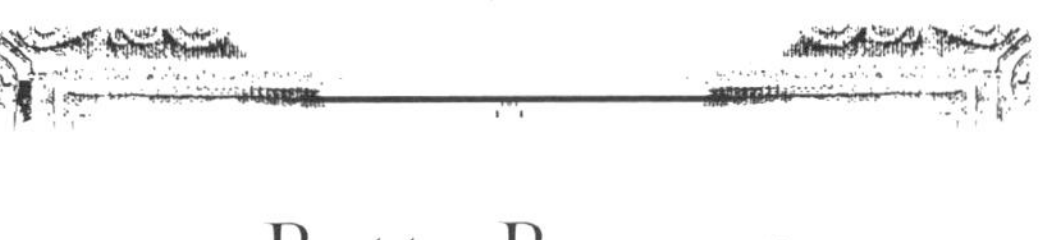

Betty Bonner

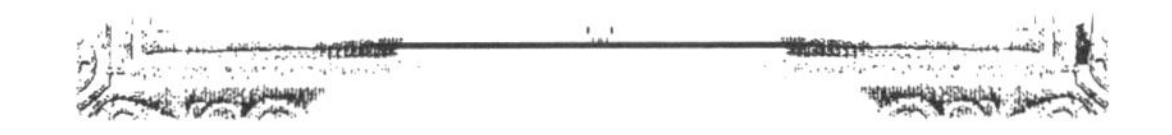

One summer in the late 1940s, Betty Bonner pointed her canoe down the San Joaquin River and didn't look back. Sitting by the river decades later, just a few miles from where she set off that day, Betty fondly recollected her youthful adventure.

I HAD A CANOE, AND PEGGY MOSGROVE and I were outdoor girls. I don't know exactly what year it was. We were old enough to be married, but we obviously didn't have much sense. Our husbands didn't want to come with us; they were both in business. Peggy and I just took off, no permission necessary. We left our husbands behind and took a trip down the river.

We put my canoe on top of Peggy's mother's car, and she dropped us at the top of the river, right where the dam is. We just got in and set off. The canoe was loaded with sleeping bags. One of us would sit up against the sleeping bags in the stern and the other would paddle. We took turns. It was fun, great fun. We took food with us, of course, and everything we thought we'd need. I remember—because I don't drink and I haven't drunk for years—but I remember we took a bottle of whiskey with us. We planned to have a drink before dinner every night. But we drank it all the first night. I really don't remember any effect from it, but we got it all over with on that first night. We drank that whole bottle.

We took a pistol with us, too. I borrowed it from a friend. We were two girls out in the wilderness; we thought we'd better have something to protect us. Neither of us had ever known how to use a pistol but somebody gave us instructions before we left. We put our sleeping bags flat on the ground and built a campfire

The Bettina, *named after the adventurous Betty Bonner. Photo by E. Z. Smith, courtesy of Kaye Bonner Cummings*

every night. We slept outside, no tent. It never rains in the summer. You never need tents out here. I've been sleeping outdoors for years.

The first night out we heard some people talking, right near us, right by the river. They couldn't have been more than twenty feet away, but there were bushes between us so we couldn't see them. We got out our pistol, put it in between our two sleeping bags, cocked it, and got all ready to shoot if we had to. The next morning we woke up, went through the bushes, and saw that the river made a sharp turn at that point. We realized that those people had been on the other side of the river the whole time. On other nights, we heard moos, we heard coyotes, we even thought we heard wolves, but we never saw any more people.

We went under all the bridges and passed Fresno, but I can't tell you exactly how far we got in those four days. I'm a dreamer, so I'd thought we were going to make it all the way down the river and shoot out into San Francisco Bay. I thought we'd be gone that long. Well, of course, we ran out of water. There just was no more canoeing to be done. I think we were just short of Mendota; that's where we had to quit. I don't remember how we contacted our husbands, but I can tell you we didn't have cell phones back then. I still don't even know what a cell phone is—I guess I'm living in the past.

I don't know what it would be like to do that trip now; it would be interesting to see, wouldn't it? I wonder what the differences would be. It's been a long time. There was no problem with camping on private property then, but it might be now. None of that entered our minds; Peggy and I just took off. She was a sport, a wonderful friend. We thought we'd had a real adventure. I suppose it wasn't really much of one, but it sure seemed like it then.

Walt Shubin

Walt Shubin farms a grape vineyard on the banks of the San Joaquin near Kerman, while also avidly fighting for restoration of the river and responsible stewardship of the land. He told his story of growing up in Madera County and shared his memories of the river when it still ran wild.

We lived way out in the boondocks, in kind of a forbidden area. It was like being on the other side of the tracks. In fact, my wife's mother told her, "Don't ever talk to anyone on the other side of Madera Avenue." We lived about three and a half miles west, "on the other side." There was hardly anyone out beyond us, probably only four or five families. Most of the roads were dirt when I was growing up. My mother finally made my father move to another house because it had a paved road, and the school buses only ran on the paved roads.

It was my older brother who introduced the river to me when I was just a little guy. He used to hike me on his bicycle. I'd sit on the handlebars, and he'd take me down there to Gravelly Ford to go fishing with him. The river there today is probably as wide as this little room, but back then it was as broad as a football field.

I think that in my time the San Joaquin River—when it was still wild—was probably one of the most beautiful rivers in the world. Wildlife was abundant with ducks and geese and big salmon runs. Whenever someone caught a salmon off Skaggs Bridge, the word was out at once, and the next day they were standing shoulder to shoulder. Before television and swimming pools and other modern distractions, the river was the place to be. From Yuba Avenue to the west—all that was under water when I was a kid. I remember going down to the river and seeing millions of ducks. We would scare them up and we couldn't even see the sky. There were so many ducks that they couldn't all fly up at one time; they would have to fly up in great flocks. I don't expect to ever see anything like that again.

When the water receded after a big flood, there were sloughs everywhere, and little lakes. When I take people down the river now and tell them what it used to be like, they look at me like I have rocks in my head, because now the river seems more like a canal. Before they built Friant Dam, the San Joaquin was a thing of beauty. I can't even exaggerate how beautiful it was. I don't have any photographic proof; no one I knew had cameras in those days. It's all just memories now.

Now I live about a mile upstream from Gravelly Ford. There used to be a Jones family that lived there, they were probably some of the original people that lived on the river. Further downstream

there was a Scribner family whose old house, an old clapboard, is still there. When I was old enough to pedal the bike myself, I'd go down and talk to old Grandma Jones. I remember once I had just read the book *Joaquín Murrieta* and I was telling her about it. "Oh," she said, "I knew Joaquin, he used to come through and trade his horses with us." I guess his horses would get real tired, so Joaquin used to stop, swap horses, and eat with the Joneses. On the way back he'd stop and trade the horses back again.

Grandma Jones told me about this one time when there was a big robbery. Joaquin Murrieta had supposedly stolen all this gold. He stopped by the Joneses' and changed horses and took off right away. Rumor had it that he buried the gold somewhere between Gravelly Ford and the Mendota Slough. Mrs. Jones used to laugh about that and say, "There's people out there been digging for twenty years for that gold." Apparently nobody ever did find it. Now that area is all developed.

Four of us friends built canoes when we were in high school. I guess we were about thirteen years old. We decided to make them as a shop project. We got a plan, and we cut the ribs and slats and covered them with canvas. We sealed them with something called banana liquid that tightened the canvas. When school was out, we ended up taking them out to my folks' barn. When the canoes were ready to go, we decided to float down to the Mendota Dam. All the other mothers consented, but I was the youngest in the pack, and my mother thought we would drown or something. I cried and complained and finally got to go.

We left at daybreak. We loaded the canoes in my older brother's old 1934 truck and he helped us take them down to the water. We took food, but we didn't have to take any water; you could drink out of the river back then. The water was still really clean. My brother said, "Now Walt, when you guys camp, make sure to get out on a sandbar." The mosquitoes were thick on the river and he said there'd always be a little breeze on the sandbar, so they wouldn't bother us. But when we went to bed that night, the salmon were running across that sandbar, and we could hardly sleep. We hadn't noticed them so much in the daytime, but I guess at night they make another run. The next morning when we could finally see them, they were making a wake that looked like a motorboat. That was a wild adventure.

I was a kid during the Great Depression, and a lot of people practically lived off the land. I used to know this guy named Emory Cauble, and I think he just about supported his whole family by spearing salmon and selling them to people in San Francisco. He used to do the same thing with ducks. He drove a panel truck, and he'd fill that truck full of ducks and drive up to San Francisco to sell them. People used to do the same thing with jackrabbits. Different seasons, different prey—ducks, jackrabbits, salmon. There was an abundance of food.

Farmers get water out of the river with really huge pumps. In fact, when you're canoeing, you have to paddle real hard to get past them or they'll suck you into the screens. By the time the river gets down to my place, it's pretty well pumped out. For fifty years, everything from about four miles below my ranch was completely dry. Two or three years ago, the Bureau of Reclamation released more water; it let about 800 cubic feet into the river for the first time in fifty years. Some friends of mine wanted to be the first to go down the river. When the water started flowing past the ranch, my son called me and said, "The water's here." We waited a couple of days to make sure it was all the way down. It was a good flow. I thought it was going to be a smooth ride, but the river is completely grown in with trees. A few miles below Skaggs Bridge was kind of like whitewater river rafting on the Kings, but instead of going over rocks we were going over trees. Several times we lost everything—hats, shoes, our cameras—everything we had in there. We still had about eight miles to go, and we were pretty fried, but we were lucky to get through that

Walt Shubin's uncle Dutch and his catch of the day, May 1937. Courtesy of Walt Shubin

Walt Shubin (walking towards cousins) just after jumping from Skaggs Bridge into the river, summer 1947. Courtesy of Walt Shubin

mess. At one point when we thought we were past the bad part, I got tangled in a tree. I had just taken off my jacket when the canoe tipped and I went crashing through the tree, into the limbs. I think if I'd had that jacket on, I would have been snagged; I'd have been a goner. It was quite a ride.

The bureau let that water down to see if more trees would grow. A small flow has been going down that part of the river ever since then, and there are trees growing there now. Unfortunately, they're growing in the middle of the channel. I'd like to see them growing on the outside, to make room for a good flow, but at that point it isn't really a river anymore. The river channel has been bulldozed and they've dumped concrete on the banks to keep it from washing out. Of course, the last flood we had went over the top of the cement and everything they had put into place flooded, just like when the river was wild.

There must be a million tires dumped upstream from me, and every time it floods they wash downstream. I would love to see this problem cleaned up. To me, as a farmer, the most important things we have are the air we breathe, the water we drink, and the food we eat. It seems like all three of them are being trampled to death. I'd like to see the whole river revived. I would love to see a parkway go all the way to the Delta. Wouldn't that be something?

I acquired my current place about twenty years ago. Most of our property is in vineyard, but there are about ten acres of bluff that go right down to the water. I don't farm that land; I just have a few head of cows to keep the grass down. Every time anyone wants to go down the river, I take them. One of my wife's relations was here this past year and I asked him if he wanted to float down the river sometime. He said, "What river?" He's lived here all his life. He said, "You mean that thing you cross over to get into Madera County?"

"To me, as a farmer, the most important things we have are the air we breathe, the water we drink, and the food we eat. It seems like all three of them are being trampled to death." —Walt Shubin. Courtesy of Gene Rose

Anyone who hasn't been down the river should go. We usually take a bottle of wine or two, a block of cheese, and some fruit. For some reason, in the beauty of the river, it's absolutely the best wine and cheese and apples you'll ever eat.

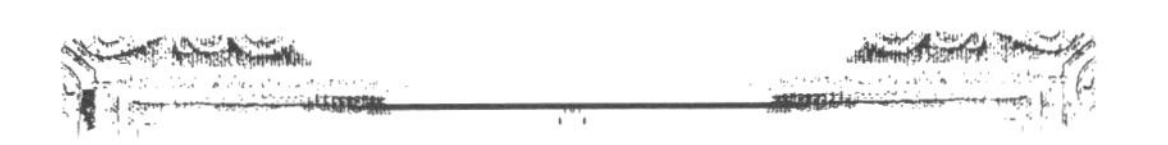

Bob Winter

Now retired, Bob Winter was a professor at Fresno City College and a science supervisor for the Fresno Unified School District. His reputation as a field biologist is known throughout the Central Valley. Bob has been leading field trips to the San Joaquin River since 1948, and anyone lucky enough to sit with him along the banks will be caught in his contagious enthusiasm for all life along the river.

I'VE BEEN COMING OUT TO THE RIVER SINCE 1930. I was in kindergarten in Heaton Elementary School when I took my first field trip. We visited the Ball Ranch, and I remember it was all very interesting to me, even then. We had punch and cookies, rode a little pony, and met the entire Ball family.

I have early recollections of coming out to fish with my dad early in the morning. We'd see Julius Trombetta, the superintendent of schools in Fresno, and Jack Savory, a coach—all hardcore fishermen. They'd all be out here real early every morning, fishing for salmon; then we'd all go off to work or school.

My parents, my brother, and I would come out every other weekend, depending on the season, and we'd float the river in a boat. We'd put the boat in at Friant or sometimes down at Skaggs Bridge and go down as far as Mendota or Firebaugh. We could catch forty or fifty big bass in a day. Typically we just released them; it was mostly for the fun of it. We just loved being out on the river.

Back then, the San Joaquin was a naturally flowing river. There was always a tremendous spring runoff from snowmelt that flushed the river and cleaned it out. Then in the fall it became a very slow, warm-running river.

I remember the gold dredging during the Depression. There were little houseboats anchored out in the river. They'd run their pumps, bring the sand up off the

"I remember the gold dredging during the Depression. There were little houseboats anchored out in the river."
—Bob Winter. Courtesy of Gene Rose

bottom, and try to get a meager amount of gold just to survive. During the hard times, that was a great way of surviving.

I enrolled in college when I came back from World War II. I was a biology student and my main professor, Dr. Leo Hadsall, had me out here teaching Boy Scouts and Boy Scout leaders starting in 1948. They were out for their merit badges and Leo didn't want me to let them off the hook too easily; he wanted me to really teach the boys. I did a whole course on nature study and we worked on the river. I always had access to the Ball Ranch; Willis Ball gave me the key and told me to bring my students out there anytime. When I became the science supervisor of Fresno Unified School District, I began bringing elementary schools out there—and I've been doing it ever since.

The river is a perfect biology classroom, but it's all mixed in with history lessons, too. For instance, some of the most spectacular trees are the native sycamores and the valley oaks, but most of those have been cut down. The giant sycamores were often cut down for buggy wheel rims. The valley oaks were cut down for furniture, flooring, and firewood. We had this terrible philosophy of taming the land by eliminating the trees and converting them into farmland. This was also a means of reducing the resources of the native peoples. Many of the foods the Native Americans ate, both the plants and the animals, were dependent on the trees. You could dominate the natives more readily if you eliminated their food sources.

There are many Indian rocks through here, lots of bedrock mortars. The Native Americans of this region were dependent upon the salmon and they allowed people from all over the valley to come get in on the huge salmon run. Lost Lake was the beginning of the spawning ground. The *Handbook of the Yokuts* was written by a man named Latta, and I knew Latta personally. He said that every bush and rock and grass out here was covered by spent salmon that had already spawned.

We used to see big clams in the river, and when my own children were younger, I'd have them dive and look for them. Those clams were one of the major food sources of the Native Americans; we've found huge shell mounds in the valley that show that the Native Americans lived on those clams to a great extent. There are little clams in there now, the Asiatic clams that were brought in for fish bait, and they've taken over everything. Some of the sloughs and channels have to be dredged because the clams impede the flow of water. But our big native clam is disappearing.

We've lost a lot of habitat from all the changes that have taken place on the river. The varieties of native shrubbery have greatly diminished. At Lost Lake Park, we used to have a native plum tree; we had several foothill shrubs and some redbud, but they've all been eliminated. Now we have introduced trees along the river: cork oaks, eucalyptus, and non-native cottonwood and pine. The latest reports on eucalyptus trees are that the pollen gums up the hummingbirds' beaks and they eventually die from it.

Yesterday I saw a big flock of white pelicans sitting down on the gravel lake. They're also disappearing because of habitat loss. We also had another rare bird here called the bank swallow. Some years we see the mountain jays down here but not very often. We also see migratory varied thrushes. There's one with a black V on her neck and she comes in for berries on the mistletoe. The other mistletoe bird is the phainopepla. The male is black with a white silver-dollar patch on each wing and a loud call.

People don't realize that cutting down trees causes many things to change. We thought the wood ducks out here were being shot by hunters, but it turns out that it wasn't the hunters at all. Wood ducks nest in hollow trees and we cut down all the hollow or dead trees and all the dead limbs, so there are no nesting places for the wood ducks. The same with the little hooded merganser. I used to see all three kinds of mergansers out here, but not lately. Another neat bird that used to be out here is the yellow-billed cuckoo. They're quite rare now. We used to have a number of Swainson's hawks that nested along the river too. Recently we've only had one pair. They've also disappeared primarily due to habitat loss.

There were some giant garter snakes on the lower reaches of the river but now they're on the endangered species list. The last one I saw was about five years ago. Many people kill rattlesnakes. They're not a major threat for folks, they rarely bite anyone, and if somebody does get bitten they don't die. The rattlesnake's venom is not primarily for protection; it's for killing and digesting its food. I show my students how to pick snakes up safely and I encourage them to carry them off the road so they aren't harmed. I've never had a student even get close to being bitten; you just have to know what you're doing.

I've been talking about these disappearing species, but there's a new colony of egrets and great blue herons just below the dam; they're nesting there now and there's a colony upstream of black-crowned night herons. We also have goldeneye ducks that we couldn't see when the river ran rapidly. We see ring-necked ducks out here quite commonly; there are a lot of little bufflehead ducks, little white guys that paddle around out here with the twinkling wings. When the male puts his white hood up they are really something to see. We've seen the bald eagles move in, along with red-shouldered hawks. The red-shouldereds trade nests with the owls. Great horned owls don't build their own nests; they're dependent upon the red-shouldered to build a house and then they come in the next year and trade off periodically. We also have long-eared owls. We had a pair right over the restroom at Lost Lake Park for several years.

In the old days by the river, particularly downstream where it got a little marshier, the mosquitoes ate you alive when it got dark. But in recent years we haven't had many mosquitoes because everything is drained, and a free-flowing river doesn't promote the growth of mosquitoes. People might see that for the better, yet the mosquitoes are on the food chain for many birds, and there are several plants that are pollinated only by mosquitoes. The male mosquito is strictly a vegetarian; it's the female that has to get a blood meal in order to lay a vital section of eggs.

We have two new birds that have taken the place of the yellow-billed magpies: crows and ravens. There never were any here when I was a child, but now I see them here all the time. They wipe out all the native species. Ravens prey on the doves, their eggs and babies. They also eat the white-throated swifts, an insect-eating bird, another friend of ours. I once held a class and we counted over four hundred ravens in one flock. They are a new bird in our area and they are nothing but trouble.

The other undesirable bird that's moved in is the double-crested cormorant. They're a very ancient bird. They destroy trees wherever they nest. Their feces are so high in waste matter that they shortly make the forest barren. Consequently, other birds have fewer trees to nest in, so they have to move on after the cormorants move in.

Of course we also have lots of koi carp and goldfish in the river. I used to see people on the dam with their children hugging a goldfish bowl. They'd been to a school fair and won three or four goldfish in some game. Thinking they should put them back into nature, the family brings them out to the river and they all say, "Bye, fishie," and off they go. We have goldfish planted everywhere; unfortunately, they've introduced foreign diseases into the water. People bring other species out in the same way. I'm fairly certain they put in some piranhas at one time. We've had snapping turtles introduced. Some of these ducks and geese out here were somebody's pets that got dumped. Feral house cats have eliminated many of the ground-nesting species and a lot of the snakes and the other little critters that are part of the food chain. We have a lot of painted turtles down in the gravel pits that people released, and they're actually reproducing. Is it good for the environment? I doubt it.

One day I saw a fellow walking from his old car down to the lake and he was carrying a bait bucket. I said to myself, "How come he doesn't have a rod and reel with him?" I thought, "I bet he's taking something down there." I rushed down just as he dumped them in. He'd brought crayfish all the way from Texas to put in our lake. He was arrested and we caught most of the crayfish. He had brought those things in innocently enough, but they're loaded with parasites that we don't have here.

Even the deer herd on the river was originally created from fawns that people found in the mountains. Somebody would see a baby all huddled up and they'd pick it up and bring it to town, not knowing that the fawn's mother was just hiding—deer are very shy creatures. People used to take them to Roeding Park, and the director of parks was a friend of the Balls. He'd bring the fawns out to Grandfather Ball, who had agreed to raise them. Finally, the Fish and Game people criticized him for having tamed deer, so he let the fences down and the deer walked away and established a herd. There's quite a herd that lives up by the dam now and we have some migrating down the shoreline. The herd is fairly healthy right now but we should probably bring in more so we can get a little interchange. We don't want the genetic basket to be all identical—that could be destructive eventually. The albinos are really unique, but people won't leave them alone. If there's a white deer, they want to shoot it. I don't know why hunters have that impulse, but a white hawk, or white anything, is going to get shot right away.

One of my favorite stories happened one afternoon when I was sitting by the river eating my lunch between classes. I was watching an old couple sitting on the bank in their camping chairs, fishing. They weren't very agile and they weren't catching any fish. There was a young guy running back and forth, catching fish constantly. This guy had three different stringers of fish hidden at different places along the bank. I finished eating my sandwich and I went over and said, "You know, I've been watching you and you're way over the limit." He said, "No, no, I've only got a few." I said, "You've got a limit there, and one there, and one over there. And I've photographed you and your car and your license plate." Which I had. So I made him an offer: "If you want to make an issue of this, I'll go to Fish and Game and I'll take the matter all the way to court. But if you want to escape today, take one limit of fish, give it to that old couple, and re-rig their lines. Give them a bubble and a fly so that they can catch a fish for themselves. Then give the other limit to somebody else and get out of here." He said he wasn't going to do it, so I said, "Well, that's up to you—you've made your choice." He grumbled but he went over and gave the old couple his fish and fixed their line. I swear, that lady's first cast caught a fish. I thought she was going to fall in the river and drown. Boy, was she happy! The fisherman flipped me the bird and went off spinning his wheels.

I'm here on the river all the time, so I try to be a good influence. I was here with my college class just yesterday. It's my fifty-fourth year of teaching and my students are scattered all over. They often tell me what it meant to them to be out here. So I guess I'm making a difference. I'd like to think so.

Ken Hohmann

Ken Hohmann told his story while leaning against a log on the dusty ground at the exact site where, at seventeen, along with over two thousand other workers, he helped build Friant Dam. More than sixty years later, he remembered every detail.

FOR SOME REASON, I DECIDED TO GRADUATE from high school a little early. So I was still seventeen years old in July of 1941, when I went to find work on the dam. Well, the union rules stipulated that "all employees shall be a minimum of 21 years," but somehow or another I was plugged in as an exception. There was one other young fellow who worked on the shift following me, and we were put in what we called the Batch Plant. That's where all the "goodies" were mixed in a very deliberate and precise fashion in order to manufacture "mud," commonly known as concrete. You need a lot of mud to make a dam the size of Friant. It was very exciting to be part of such a big project—building a dam. I was paid a dollar an hour. I have no idea what my job title was; nobody ever told me. Basically, I was a dumper. I made sure there was a can down below me; I would flip a lever and throw four yards of mud down the chute.

The whole production was quite a scene. There were four of these cans on every donkey train. The cans were huge—six and a half feet in diameter, ten feet high, and weighing nine tons. Each one handled four cubic yards of concrete. We made a batch every nine and a half minutes—an awful lot of mud. The high-time pour was about six thousand cubic yards in twenty-four hours. You can imagine how many cans had to be dumped to fill one of those blocks. The crane operator couldn't always see what he was doing, so they had signal boys, whistle boys, down next to the cans. They'd give instructions to the crane operator about what to do with these humungous cans. There were about five men in high rubber boots that operated in this mud. They had to hold onto a huge handle and dump the mud in a controlled fashion; they had to move it back and forth, and the whistle boy helped them move the cans around. Then they would vibrate the blocks to eliminate any voids, or air spaces. After the pour was completed and they got it up to the lift, they'd smooth it off and let it cure and move on to other blocks. There were always several blocks operating at one time. It was like a three-ring circus. I was in awe of the spectacle.

When I first came up here to Friant, with no car and no money, I was assigned to the dormitory. I had my little fake alligator suitcase and I guess my folks had bought me some suntan pants, because I remember I had a couple sets of

"You need a lot of mud to make a dam the size of Friant. It was very exciting to be part of such a big project."—*Ken Hohmann. Courtesy of Fresno County Library*

suntans and one pair of work shoes. I was wearing glasses at the time, and wearing glasses while working in the batch plant wasn't fun and games. There was cement everywhere. At the end of the shift, I could take my trousers off and stand them in the corner. I couldn't even see through my glasses.

The dormitories looked something like army barracks, except there were only two men per room, like a one-room apartment. I shared a room with another young fella, though he was maybe twenty-five, which is pretty old when you're only seventeen. I don't remember his name, or even what he did exactly, but he worked *on* the dam, whereas I worked *off* the dam. There were two dormitories with toilets and showers, gang-style, at the end of each building. Unfortunately, the aggregate conveyor belt was right next to our room and it made a horrible racket. Other than that, it was pretty good living.

We were thirty miles from town and we'd already been selected to work on the dam, so the contractors really wanted to keep us there. They didn't want to do anything that would upset us, especially in terms of food. So there was beaucoup food! Not only in volume but in choices. For breakfast there was every kind of egg you could imagine, scrambled or easy-over; they offered everything but soufflés. Cold cereal, hot cereal. Milk, coffee, juices. All family-style, served on a huge plate. If we worked the day shift we picked up our lunches—a brown bag with a couple of big sandwiches and an orange—and we'd take that with us to the job site. Dinner could be anything from pork chops to meat loaf to chicken—again, plenty of food. I couldn't believe how fortunate I was. In the contract, it said that we could not be charged more than a dollar and twenty-five cents a week for our board and room. That's how much I paid—for all that great food and a bunk and a bathroom.

Excavation in progress. Courtesy of Fresno County

"When I first came up here to Friant, with no car and no money, I was assigned to the dormitory."—Ken Hohmann. Courtesy of Fresno County Library

There were sixteen hundred men working on Friant Dam, on four shifts per day. It was that fourth shift that was trouble. The three regular shifts were the day shift, the swing shift, and the graveyard shift. But we didn't work seven days a week, only five, so somebody had to come in and pick up the extra two days. That was done by a relief shift. Now *that* was fun and games. Let's say you finished your regular assignment; you finished your month and you were getting off work at midnight. However, if your next month's shift was to be the relief shift, you would have to come back at eight the next morning. You would have only a few hours off, then you would work the swing shift. You'd do that for two days, then the next two days you would work the graveyard shift. The day after that you would have to work the day shift. You didn't know if you were coming or going and you did this for four weeks straight. That was rough.

For a young fella, there was sure a lot to be interested in. One time the hopper operator, who had a drinking problem, was apparently thinking about something else during an operation. He didn't watch the timer and he pushed the sand valve onto the top of the cobblestones. What a mess that was!

There were all kinds of rumors flying around. There were rumors that we might find some gold in the river and that the Bureau of Reclamation was going to keep all the gold. Another rumor claimed that there was enough gold taken out of the river to pay for the whole dam. But what I understand now is that the original contract stipulated that half the income from the gold went to the contractor and the other half went to the government. I never did hear how much they found, but I bet there was plenty.

By the time I got there in 1941, everybody knew we were going to war. We didn't know when, but we knew we were going. Now, to build the spillway sections of the dam, huge steel drums were needed, each one eighteen feet high and one hundred feet long. By the time I got there, the Friant Dam project didn't have any priority for steel. On the other hand, Shasta Dam had top priority because we needed its electricity for the war effort. We had men on the Friant job who were jumping ship to go work on Shasta. This meant that Friant Dam wasn't operational until 1946 or '47. We had to wait for the war to end to finish it and then it took a while for everything to come back into full steam.

Friant Dam. Courtesy of Fresno County Library

It's funny what you remember when you're a young fellow. There's one story I remember like it was yesterday. I told you that I didn't have a car. I had lost a tooth and I needed a ride into town. I went down to the administration room, to the paymaster, and out came Harvey Slocum, the superintendent and famous dam builder. He just happened to walk through the office as I was asking about getting a ride and he said, "I'm going into town, I'll give you a ride." So—I got to ride with the great Harvey Slocum. I finally got to meet the man himself. He reminded me of George C. Scott; he was that kind of guy. Now I'm quite a talker and I asked lots of questions. His car was a beige Buick Century sport coupe with twin fender wells. It had spotlights and fog lights. The car had been given to him because construction on the Grand Coulee Dam had come in ahead of schedule and all the top guys had gotten a bonus. The fellas that worked under Harvey Slocum all knew the Friant job was coming up pretty soon, so they chipped in and bought this special order Buick with a Roadmaster engine. Harvey was known—and probably pretty proud of it, from the way he talked—to have a heavy foot. He liked to play golf, and the day I rode with him, he was heading for a golf game at Del Monte, near Monterey. He told me how he'd get over there in two hours at ninety miles an hour in that big Buick. He'd pass anything that got in his way. I guess every highway patrolman between here and the moon knew about him.

Out of the blue, just before Mr. Slocum dropped me off at the dentist, he asked me how many work boots I had. I said I only had one pair. He said, "No, no, no, no. You have to have one pair for every workday of the week. You need five pairs of boots." He was quite a guy. Meeting him was an honor for me. But I never did have five pairs of boots, that's for sure.

2327 – SAN JOAQUIN RIVER. CALIFORNIA, FROM THE HEIGHTS NEAR FRESNO.

Bill Loudermilk

Fisheries specialist Bill Loudermilk retired from the California Department of Fish and Game, where he served as regional manager, in December 2008. His story holds a wealth of wisdom and is told with great affection and concern for the future of the San Joaquin River.

I WAS BORN IN 1949. We lived in Isleton, a small town in the Delta. It's near where the Sacramento and San Joaquin Rivers come together. My father worked and my mother was a stay-at-home mom, but she loved to fish. I would come home from school and my mother would have the family station wagon all packed up with a picnic dinner and fishing rods. She would sweep us off to the river, my two sisters and me, for an evening of fishing. We had an old wooden boat that my dad would row across the San Joaquin River at the mouth of the Mokelumne River down on the Delta and we'd fish a few places where there were a lot of largemouth bass. My whole family loved to fish and camp and hunt. I can remember the first fish I ever caught; it was in Yosemite on the Merced River, one of the tributaries to the San Joaquin River. I think I still have some 8mm footage of me when I was probably all of three or four. I didn't know how to use the reel yet so I just backed up away from the river and dragged the poor fish out.

My love for fishing is probably what helped me to focus on a career. I wasn't a particularly wonderful college student. At the time, the draft was in full swing and I spent some time in the army and about a year in Vietnam. I realized that I really needed to do something with myself, so I reconsidered my priorities and ended up with a degree in environmental science. After I graduated, I worked actively to get myself on all sorts of hiring lists with federal and state governments, resulting in a call from the National Marine Fisheries Service with an offer for a fisheries technician position. I had hoped for a position in a freshwater fishery, because that was my main interest, but a marine environment is really a phenomenal place to work. After three months there I was offered a position with the Department of Fish and Game as a scientist at Big Bear Lake in Southern California, in the San Bernardino Mountains. Oh boy! An opportunity to go back to freshwater fisheries work. I transferred later into a position on the Lower Colorado River, and I also worked as a fisheries biologist in the Owens and Walker River basins of the eastern Sierra.

I was promoted to Fresno in 1983. When I got my first look at the San Joaquin River, I said, "Good God, what

Posstcard view of the pre-dam San Joaquin River. Courtesy of San Joaquin River Parkway

happened?" This midsection of the San Joaquin Valley's namesake river had but a dribble of water remaining—or was entirely dry. The adjacent land was farmed right to the river's edge, leaving only a thread of riparian forest. Only a small fraction of the fish and wildlife species diversity remained, and those remaining populations were very low. Adding to this insult, the public was using cars, household appliances, and plenty of vehicle tires for bank stabilization. There was active dumping of household and agricultural refuse.

I had grown up on the lower end of the river, where there was plenty of water, and I had no idea what the upper river was like. As a scientist, I perceived that as a travesty and I channeled my outrage by taking a more active role in the American Fisheries Society. I started talking with some compatriots on the environmental committee of the society. I asked, "Isn't there something we can do about the San Joaquin River?" I was president of the society at the time when we and other organizations ponied up some of the first few dollars that helped support the Natural Resources Defense Council (NRDC) to file a complaint on the San Joaquin River. We helped get the whole lawsuit rolling. And now, finally, we've won. We're going to see the river restored.

The law that was the basis of the NRDC complaint had been on the books for a long, long time, since the 1940s. It's the Fish and Game Code Section 5937 that requires the owners and operators of dams in California to release enough water to keep fish alive and in good health. Basically, that's been the law for many years, but it wasn't until there were some substantial societal changes that anybody was ready to enforce it. In 2006 a settlement agreement was finally signed that will restore the water flows and salmon all the way up to the Friant dam.

The main focus of this new program will be on restoring what is known as a "spring run" of salmon. Prior to the construction of the dams, the spring run was predominant. The salmon migrated upstream on the snowmelt. The rivers flowed heavily in the spring and into early summer and these Chinook salmon moved up out of the ocean, through the Delta, and migrated on those big spring runoff events.

To restore the spring run, we are actually going to have to bring some spring-run fish from other tributaries into the Sacramento–San Joaquin system; fortunately, we have a few remnant stocks on the Sacramento side of the Central Valley. Initially we will have to introduce some fish; we will have to rely on ample numbers and a cautious hatchery program to assure that natural reproduction is started properly. As the release of flows that are required under the NRDC settlement agreement begins, we will start to restore the spawning areas, eliminate some of the blockages to migration, and screen some necessary water diversions so the young fish are not diverted into agricultural fields. We are going to incrementally bring new spring-run fish in each year, get reproductive populations started here, and over the years more and more of these juvenile fish will be able to migrate safely to the ocean, survive and return as adults to this river, produce their young, and perpetuate the restored population.

In the case of the fall run, we still do have viable runs in the three northern tributaries but Fish and Game has been operating a physical barrier on the San Joaquin River—the Hills Ferry Barrier. We physically preclude fall-run salmon from migrating any further up the San Joaquin River than the confluence with the Merced River. We have had to do that because there are so many migratory blockages, plus the absence of water, that if we allowed those fall-run fish to migrate any further than the Merced River they would be lost. When we start adding water, by virtue of the settlement agreement, there will suddenly be a connection again to some historic spawning areas. We may not actually have to use the hatchery to restore the fall run.

When salmon live in salt water and suddenly come into fresh water, they need a transition period. Physiologically it is tough on them, so they stage on the tides, move back and forth, make daily forays into fresher water, and pretty soon they have adapted enough so that they can come farther upstream. By mid-September they move up out of the Delta and into the river systems. The leading edge of the spawning migrations starts to show up below the dams on the Merced, Tuolumne, and Stanislaus Rivers. By the first of November we start to see some spawning activities; the females begin digging their nests in the spawning gravel. Most of the eggs are deposited in the gravel by mid-to-late November. Come January, the eggs have been incubating in the gravel; we start to see some of the young fish that have hatched but they are still absorbing the yolk sac that was conveyed from their mother. Eventually they have to come up out of the gravel and take up life as free-swimming fish. They are very small, so they cannot deal with fast-moving currents; they tend to find the slow-moving areas of the river. From about January through April and May they go from a very small, weak-swimming fish to a fish that is maybe three to four inches long. They reach the life stage that is referred to as a smolt, and that is when they migrate to the ocean. It is quite a process and we are going to see some of it happening right here.

It is exciting to help put the river back into a more healthy condition. It is interesting that it took more than sixteen years of litigation to get us to this point of the restoration effort. It has really been amazing to me to see the dramatic change in the social expectations of people as they revalue the assets associated with a viable river, one that has been restored to its original, natural state. As California grows and we are able to provide the public with opportunities to enjoy the state's rivers again, we hope to instill those same values that I learned fishing as a kid with my family, an appreciation for the value of our natural world. Then, when future generations are the decision makers, they will make sure to protect this world in perpetuity. That is my hope.

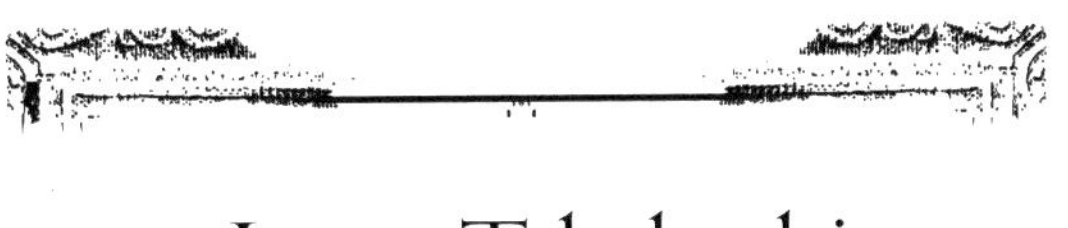

Irene Takahashi

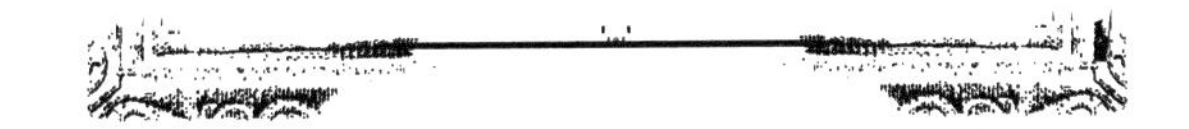

Irene Takahashi married into one of the most well-respected farming families in the San Joaquin Valley. She told her river story in north Fresno, at the corner of Copper and Millbrook, where she'd spent a hot June morning selling boysenberries at one of the family's roadside fruit stands.

BEFORE COMING TO LIVE HERE, I was like a square in a round world. I'm American born, second generation, but both of my parents came from Japan. I grew up in San Diego, but I had never felt like I fit with life in the city. When I married into the Takahashi family and my husband, Ted, started farming over here by the river, it was the life I needed.

We started farming on the San Joaquin River in the early 1970s. That's when the gravel company asked us to come farm there, where the pits used to be. Ted knew the soil was like silt, river bottom soil, very good for what we wanted to raise. River bottom soil is rich, very rich.

When we came out here, it was all still very pristine, even into the late eighties. There was a lot more wildlife then. The beavers were such mischievous critters; and there were deer and a lot of raccoons and muskrats. And there's a lot of history by this river—lots of history and lots of—you know—spirits.

We moved into a house that Mr. Gibson had built right on the river, a beautiful tile-roofed house. He'd built that house for his bride, but Mr. Gibson passed away before they got to live there together. His bride never got to live there at all.

Can I tell you a story? Can I tell you what I mean about spirits? There was a lady who more or less lived with me then; she did the housework. We were close. We had some kind of relationship; I don't know what you'd call it—but we *sensed* things. I came home one day and she was at the kitchen window; she had all of the doors locked. Her head was down and she was scrubbing the same spot over and over. I said, "Emily, let me in." She shook her head. I finally convinced her to let me in. She said, "Irene—there's something in this house." I said, "What happened to you?" She told me she was putting the linen in the hall closet and she looked up and saw something with a veil on, looking at her—just a head. I said, "That's funny, Emily, because the same thing happened to me." I'd been in the side bedroom one night and I could see something moving, like leaves, in the same hallway. It was like a veil, just floating. I'd never told anybody because they would think I was nuts. I hadn't told Emily either and I never imagine things. It just floated there for a

Irene Takahashi's father-in-law, Yoshibei Takahashi, tends his farm along with his faithful friend Sammy in this 1937 photograph.
Courtesy of Lorraine Takahashi

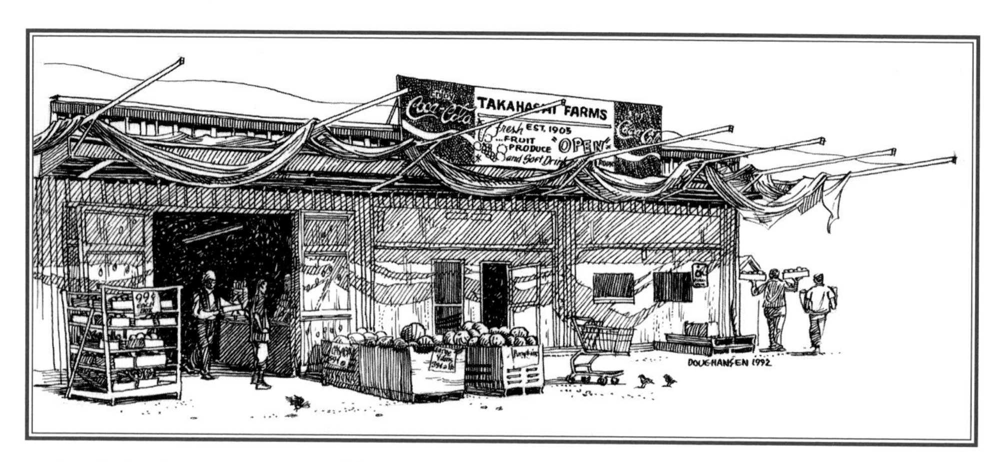

A Clovis landmark since 1962. Courtesy of the Fresno Bee

while. There was nothing frightening about it; it didn't scare me. There was actually a kind of loneliness about it. I definitely was not afraid, and Emily said she wasn't either. I don't believe in ghosts, but this was a presence. We thought it was trying to tell us something. I thought maybe it was the spirit of Mr. Gibson's bride, so I went to St. Agnes Church and got a statue of a saint. I put it down in front of Emily to ease her apprehension and said, "Will you still work for me?"

There were still a lot of native people around here when we first moved here and I liked that. I feel comfortable with the Indians. They communicate in a very majestic way. When we started farming, it was with mostly Indian workers, not the Hispanic farmworkers yet. We had an Indian lady; she was their boss. She was real silent. People would rant and rave to her and she would just nod. She was very honorable in dealing with us and the workers.

When we first moved out here, one Indian man told me, "There's a fox with a white face that lives out here and I honor that fox." Later, one of my sons ran into a hunter who had a fox cornered, a fox with a white face. The guy was going to kill it but my son said, "No, this fox is honorable with the Indian people; his spirit is alive. You can't shoot him." My son understands about those things too. I think the Indians have a good history; that's part of what makes this river a magical place. I have deep respect for the Indian people here and the Indians have deep respect for this river. I have respect for it, too.

There's a special feeling out here by the river. Bud Rank once told me that his dad was never a churchgoer but that he'd go out to Rank Island every Sunday. It was like a cathedral out there on that island; that was his church. For a while we planted boysenberries on the island. It is so beautiful, like another world, with big tall trees, quiet and peaceful. There is an enchantment out here. The Riverview Ranch House is another special place on the river. There is an old barn still standing out there where white owls live. There are a lot of owls in that barn. One of our farmworkers saw a white owl fly out of the barn and he said it became a woman, a woman in white, maybe an angel. That man went right back to Mexico.

Joaquin Murrieta was around this area and I've heard lots of stories about him. All of these stories should be preserved because this river is unique. There is a lot of history out here, a lot of spirit. I know I'm getting off the subject, but that is how it is up here. There's magic all around.

Dale Mitchell

Dale Mitchell was the aquatics program manager with the Department of Fish and Game at the time he told his story. Recently retired, he still plans to volunteer his vast technical knowledge and historical expertise to help in the San Joaquin River's recovery.

I GREW UP IN A DEPARTMENT OF Fish and Game family, so I was what you might call a fish hatchery brat. We moved all around the state but when I was about seven years old, which would have been back in 1957, we moved to the hatchery at Friant. My dad was a manager there and I spent the remainder of my growing-up years right on the edge of the San Joaquin River. It was the primary playground for me and all of the other kids in Friant; the river was a central part of our life. It was the place where we sneaked off, where we went to do things that we didn't want our parents to know about. All kinds of kid stuff.

The people who lived in Friant at that time were a diverse group. There were families that had stayed over after the construction of the dam, people who had come from Oklahoma and Texas for the work and then stayed on. We had the remaining Mono Indian tribe that lived in the area as well. Prior to Friant, I had lived in Elk Grove, and to me an Indian was someone you read about in a book. When I came to Friant and started going to school, it took a while for me to register that these were really Indians, that Indians weren't just people who lived someplace else. These were my friends; we played baseball after school and all those things. We all mixed with the Mono kids and their families. I learned quite a bit because the Indians have a huge cultural memory in their lives. They taught us some of their skills. I learned how to pound acorn in stone mortars next to the river. We watched them make baskets. We ate the flat bread they made from the acorn meal. I didn't really think it tasted very good, but they thought it was wonderful, and I never told them I thought any differently. It was fun to be a part of it all.

There was a line between school districts right at the edge of Friant. Some of the kids who went to elementary school with me, like Warren Ball, were on the other side of the line, so they went to Clovis High School. Most of us went to Sierra High, in the foothills. When I went off to high school there, the Mono Indians that I had grown up with were mixed with the various Yokuts tribes from around the foothills. That was another revelation for me, learning a little bit more about

"We all mixed with the Mono kids and their families. I learned quite a bit because the Indians have a huge cultural memory in their lives." —*Dale Mitchell. Courtesy of Fresno County Library*

the differences between the tribes, as well as their similarities and the intermixing between them. At the time I wasn't very interested in history; I don't think most kids are at that age, but as I grew older, those details started to fit together like pieces in a puzzle. I now have an understanding of some things that didn't make sense to me then.

As a kid, I always had an aquarium in my bedroom, with hardhead and squawfish and all those native minnows, but I don't remember that I made a conscious decision to join Fish and Game. It was just a natural evolution that happened, not only for me, but for a lot of the other kids that grew up on fish hatcheries. I went to work for the department, first as a seasonal aide in the sixties; then after I finished up my degree at Fresno State, in the early seventies, I went to work as a biologist. I worked at Millerton Lake. I was involved in the revisions of the Kerckhoff project and in some upstream projects on the San Joaquin River, the Big Creek project and power licenses.

I started to broaden my understanding of the river. When most people in Fresno think of the San Joaquin River, they think of the area from Friant Dam downstream. It really is part of a larger watershed and I don't think very many people understand how it all fits together, the many different reservoirs and water operations. It's a very complex system.

I've spent most of my life trying to understand the intricacies of the San Joaquin River, and it's not an easy system to grasp. In my personal research, I've started to understand that once something gets written down it becomes a part of history, whether it's correct or incorrect. There's an awful lot that has been written about fish that doesn't make sense when you consider the biological history of the animals and what must have happened to them geologically. People have either misidentified fish, or misidentified places; they described things without the appropriate expertise. I've set out to try to sort out some of those kinds of historical messes. I've started with Native American cultural information and put that together with what we know about the natural histories of the fish. I've looked back at the old explorers, all their diaries and letters. I've been trying to digest all that with a focus on what occurred with the aquatic resources. It's been fascinating; I have a room in my house that's literally full of historical material.

We know that the San Joaquin River was a drawing point for tribes from all over the valley and that there was a mutual truce on the river during the salmon runs. These were tribes that were at war at other times, like the Chowchilla and the Dumna Yokuts from here, and the Monache Mono from up the hill. They fought the rest of the time but not during the salmon runs. They camped together along the river, shared and traded resources, fished for salmon, and pounded acorns together. The river was an attractant because it had a huge protein source every year, so it remained a peaceful, neutral territory.

The San Joaquin River's complexity poses both problems and opportunities in terms of restoration, and now that restoration has been deemed official, it's an exciting time. I can try to describe the process to the extent that anybody can. There are a number of things that have to happen in order to hit the target of restoring naturally sustained salmon in this river. Establishing a fully self-sustaining population of salmon is a huge task, much larger than people realize. To some people, the settlement agreement is the end of the fight, but to others it is just the starting point.

First, you have to be able to get the fish up and down the river, and getting fish up and down a dry river is no small task. Channels will have to be constructed. Bypass structures will have to be created. Each step in this process involves dealing

"They taught us some of their skills. I learned how to pound acorn in stone mortars next to the river." —Dale Mitchell. Courtesy of Fresno County Library

with landowners; legal agreements will have to be established. Ultimately, we'll have to make decisions about which fish to use for salmon restoration; we'll have to obtain the stock and start a biological program. Before that, however, we have years of just working out access and the physical habitat problems. The general design will have to be an adaptive program. We're going to continue to get feedback as we gather scientific information and we'll hear from different parties in the public. We'll keep tweaking and changing it as we go. It's a long-term project.

Many of the things that some of us are emotionally attached to on the river—places where we've taken walks, places where we've fished and picnicked—will have to be modified as we construct the spawning beds. We'll have to straighten out areas of the river where there are multiple channels. We'll have to isolate the gravel ponds that are growing populations of fish that will prey upon juvenile salmon. We'll have to wait for the vegetation to grow back to make it look natural again. People generally react adversely when they see a tractor operating in a river, but we can't recondition these spawning gravels without tractors.

The key is to not allow the restoration to devastate the economy of the region. That means that we have to figure out how to use water more than once in order to release enough for the restoration to be successful. That's the bottom line. The whole ecosystem of the river depends on how the water is managed. We can't use the water only for agriculture. If we take the water, deliver it by gravity out to somebody's fields, then let it percolate into the ground, we've made two uses of it. But consider other options. After the water has already served its purpose in the restoration of the river, what if you pumped the water, moved it to an agricultural field, then put it back into the groundwater—and what if you produced the energy for that pump from hydroelectric power generated by the water itself? These are the types of systems that the water management group in this restoration is looking at.

Throughout my life I have seen politics as being at odds with almost everything I wanted to see happen in my life, but after watching a number of issues work out over long periods of time, I've learned that politics really can be a balancing agent. If you're going to restore a river, you have to ensure that it doesn't economically devastate or blight an area, or disadvantage groups of people. Politics can be a good thing; it can be the philosophical process that binds people together. Skeptics are important to this process because they ensure that the project is solidly grounded before spending public money. Skeptics may be one of the more important constructive forces because they force us to do the job right. As technical people, we must understand that we can't spend public money and then just tell everybody, "There, the restoration is done." The public needs to see and understand this river and they must have access to it. We all should be focused on how we can generate tangible benefits and deliver them to the people.

I watched thirteen deer walk across that little narrow metal footbridge on the Madera Canal one morning, crossing between the handrails. The predators that prey on deer are down on the river, and they are part of the ecosystem too. In the area by the Ball Ranch, we've had a number of accounts of mountain lion sightings. The whole ecosystem is in constant motion on the river. It's a matter of recognizing its connection to so many other vital parts of life and not just seeing the river as a thing on its own.

There is an old saying about how you can't repair a watch if you don't keep all the pieces. Keeping all these pieces has to be a focal point in terms of the restoration. The public sees large amounts of money being spent on the restoration of the river bottom, but we can't leave any important piece out. So we have to decide: is this important to us? Do we commit ourselves to maintaining the whole package? Personally, I think it's a very workable concept and well worth the expense. These are individual choices, but I do think that a hundred years from now, our whole society will be judged by whether or not we saved our rivers.

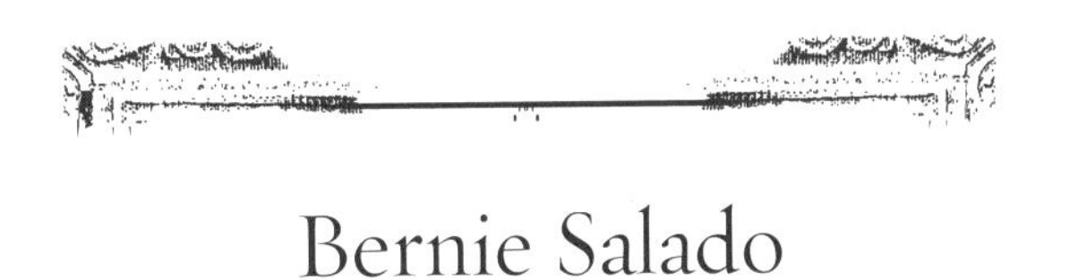

Bernie Salado

Bernie Salado, a native of Mexico, grew up on the property directly behind the restored Riverview Ranch House that now serves as an educational center for the Parkway Trust. Years later, he sat in the cool shade of the same camphor tree he'd swung from as a child and shared his memories of growing up by the San Joaquin River.

WE LANDED HERE IN JUNE OF 1968. We came from Mexico. My dad brought his six kids and a wife. Imagine, I came from an area where I'd never seen a fruit tree; I'd never seen a car or a plane. We flew into Fresno at night. The next day we woke up at this place and my dad took us to that pasture out there; all of those fields were pastures then. There was a vineyard right there, and on this side it was all fruit trees. My dad was irrigating that morning and I'd never seen water come out of the ground like that. I was fascinated by the sight. He brought me into the orchards and I saw my first fruit tree. We picked some fruit; it was like the Garden of Eden. It was paradise. Coming from a life with no electricity, no running water—suddenly here we are on this ranch by the river. It was like the Wizard of Oz, you know, where everything is black and white, then you wake up one day and suddenly it's all in color. It was absolutely a paradise.

A lot of the people from my hometown in Mexico lived around this area; they worked for the Cobbs, the Coombs, and the Ranks. Somehow a lot of us ended up here. My father and one of his cousins, who still lives down the street here, came here in the early forties as braceros. That was what they called the work permit program back then. They came and worked in Fresno; then my dad went back to Mexico, got married, had his children, and one day realized that it was not working out in Mexico, that it wasn't what he wanted for his family. He called his cousin to see how it was going here, and the cousin said, "Yeah, come on up." My father set up everything and arranged for us to come behind him, right away. It takes just one person; then they tell another person; that's how so many people from my town migrated up here. Like so many other immigrants—it happened with the Armenians—one family came and saw that the grapes were good, and so more came—same with the Japanese and Chinese. That's how it happens.

I was eight years old when we came here. My dad worked for the Rank family; he broke horses for them. When we first came, my dad was making a dollar an hour—with six kids! But we had goats and pigs and cows and chickens, and we had a giant garden over there. That's basically how we lived. It was great. We never lacked anything. You don't know you're poor until somebody tells you that, so we didn't know. There was always food on the table. I remember going for groceries and getting a hundred-pound bag of beans, a hundred-pound bag of rice, and a hundred pounds of potatoes. Everything else was here—eggs and meat and fruits and

"My dad was irrigating that morning and I'd never seen water come out of the ground like that."—Bernie Salado. Courtesy of Fresno County Library

"I think back on it, how fortunate we were. God, what a life. Of all the places in the world, we ended up right here."—*Bernie Salado. Courtesy of San Joaquin River Parkway*

vegetables. And of course, fish. We weren't a real fish family when we first came, because the region we come from in Jalisco is in the middle of Mexico, high desert, very arid; there weren't any fish there. So my mother had to learn to cook them, because there was always free fish in abundance. I remember just throwing a line in the river and jiggling it, just one quick jiggle and we'd catch fish all day long. Trout, bass, suckers.

One of the best things about living here as kids was that we could come home and swim. We'd take the horses and go straight down to the river. I was always amazed, because we'd drop off kids at the estates with their mansions and pools and nice cars—we didn't have any of that—but these kids would ride their bikes all the way out here, to spend time with us. We'd have tons of kids come over. I was like, "Hey, you've got air conditioning, you've got pools." I didn't understand then. Of course, now I know why they wanted to be here, but back then this was just my norm.

We were so secluded out here, but at the same time, on Saturdays people would show up out of nowhere and we'd have a big lunch. My mom would send us out to get eggs. The Takahashis planted watermelons farther down, so my mother would send us to get some watermelons or some other fruit. We had access to this three-wheeled little mail cart; we'd drive it down to Takahashis' and load it up with fruit and bring it back to the party.

Everything was spontaneous. We used to go down to the Ball Ranch and float down the river to our house. We hung ropes in the trees and swung and jumped into the river. We never said, "What do we do?" or "We're bored." We were never bored. My dad broke horses, so we'd go riding into the river. As we rode, we saw deer. One time we saw a little bear by the river. It was remote down there by the islands. I remember riding by the river with my dad and thinking, *This is it, this is heaven*. Those were some of the best times I remember with my dad, riding alone with him down by the river.

We had horses and pigs in there, we watched cows and horses have babies. One time, one of our sows had too many piglets and she started eating them. As kids, we were freaking out, we were like: what is she doing? My dad explained that she knew that she couldn't take care of so many. She had like eighteen. So we took six away from her and started raising them in the house. Before you knew it, we were watching TV with them. Obviously, my mother didn't like it that much. We'd start getting very attached to these animals, but we still knew about survival. We knew that these animals had to be butchered at one time or another.

I'm still living on the river, about ten miles from here. I see deer all the time. I've seen mountain lions pounce right through my yard in the early morning. I've seen the herd of white deer that live by the river. There are some deer trails that lead from my house right down to the river. There are piles of bones along there. They're deer trails, but the mountain lions know that, so they've had a few meals there.

The San Joaquin River has been such a big part of my life that I'm working now on a photo project about it. I'm a photographer by trade. I could see a lot of changes coming, so I thought I'd better do something—even if it's just for myself—to document the river before it changes completely, to show what it is like now. Everybody wants a piece of the river, and I know that something's going to happen. I understand the economic implications, but at the same time, this area is such a gem.

Our life here at this ranch was a very special time for my family. My parents are still living; I have five sisters and they're all still around. Every time we talk about living by the river, it's about very fond memories. Basically, it was a peaceful time in our lives, before all the kids got bigger and scattered all about. I think back on it, how fortunate we were. God, what a life. Of all the places in the world, we ended up right here. I'm so glad they saved this old house. They could have just bulldozed it and nobody would ever have known about it except for a few people like me.

Lloyd Carter

As a journalist, Lloyd Carter has covered water issues in the Central Valley for the last forty years, and he offers a unique perspective on the complex political and social history of the San Joaquin River.

I HAVE BEEN INVOLVED WITH THE San Joaquin River all my life. I was born and raised in Fresno in the 1950s, and I swam in the river as a kid many times. I had a good neighbor friend whose family later moved to North Fork. We used to go up to Mammoth Pool, which is the first storage reservoir on the upper San Joaquin River. That was great fun. I've got family history on the river too. My grandfather came here as a Danish immigrant in 1912, and he pitchforked salmon out of the San Joaquin. I think almost everyone who was born and raised around here has a grandparent, or even a great-grandparent, who remembers what the river was like before the dam was built. In my lifetime, of course, the river has always been dammed, and I really didn't know the history of the area at all while I was growing up. I went through the Fresno public school system and we learned about the Indians and the acorns. I remember that part, but I don't remember ever learning that the San Joaquin River dries up for fifty miles; I learned later, though.

I graduated from Fresno State with a degree in journalism, and in 1969 I went to work for United Press International, one of the two major wire services in those days. One of the very first stories I covered was the dedication of the A. D. Edmonston Pumping Plant at the base of the Grapevine, with Governor Ronald Reagan presiding. The plant pumps water from the state aqueduct in Northern

The San Joaquin was once a great fishery of sturgeon. Courtesy of Gene Rose

California over the Tehachapi Mountains and into Southern California. It is one of the great engineering feats in world history, truly an eighth wonder of the world: tons of concrete, miles of canal, and gigantic turbines built to lift an enormous amount of water over the mountains. As a young, wet-behind-the-ears reporter I was pretty much clueless. I see now that I was easily manipulated to believe in the "engineering marvel" story, which was actually a project to steal Northern California's water. At that time, though, it did not raise any larger questions in my mind.

I spent fifteen years as a newsman, and I wrote many stories about life in the San Joaquin Valley. I wrote many agriculture stories, but I didn't write about water. Then in the early 1980s I got involved in the Kesterson story. Kesterson was a wildlife refuge about two miles from the San Joaquin River that had become a repository for drainage waters coming off the Westlands Water District. In the spring of 1983, federal scientists discovered that many of the birds nesting there had deformed embryos. Birds were born without wings, without feet, without eyes, their brains protruding from their skulls. It was terrible, grotesque.

The federal Bureau of Reclamation was responsible for delivering water to Westlands, and they claimed that the problem at Kesterson was only a minor engineering flaw that could easily be corrected. Because of the unique soil chemistry on the west side of the valley, there are subterranean layers of clay. When this soil is irrigated, the water percolates down into the clay and backs up into the root zone. To deal with this problem, farmers have to bury perforated pipe; the shallow groundwater trickles into the pipes and then is pumped to a ditch at the low end of the field. At that point, they send the toxins to "Away Land," which is a euphemism for "someplace else." Fifty years ago, when the bureau devised this drainage plan, the goal was to pump the water to the San Joaquin River's terminus in the Delta, at a place called Chipps Island. The theory was that it would be diluted by cleaner water and then flushed safely out to sea, but Bay Area residents thought this was a horrible idea and they blocked the plan. The bureau then came up with a stopgap measure to take the polluted water, confine it to ponds, and let it evaporate. The clean vapor goes up, and the toxic elements stay down in the pond. That's what they did at Kesterson. They called this polluted water "summer water" or "agricultural return flow," and they actually sold this idea of dumping toxic water into Kesterson as a "wildlife benefit." Of course, the issue blew up in their faces, and that's when I got seriously involved in water issues.

I met some fascinating people through the Kesterson saga. Jim and Karen Klaus were Stanford-educated ranchers, with PhDs, who had bought a cattle ranch adjacent to Kesterson. Their cattle started dying and they discovered that water was leaking from the Kesterson ponds onto their lower-lying land. They were the first whistleblowers, although every cattle ranch around Kesterson had dying cattle. The Klauses came to me, and I began writing about the problem in the early spring of 1984. I put together a five-part series that ran on the front pages of papers all over the country. The CBS program *60 Minutes* even did a big piece on it and the *New York Times* ran it on the front page. In February 1985 the State Water Board ordered a cleanup. They told the Bureau of Reclamation to clean up the mess or shut it down. A month later, at a historic congressional subcommittee meeting in Los Banos, the secretary of the Interior ordered Kesterson closed. All this because Jim and Karen Klaus filed a complaint—two citizens taking on the United States government. It was extremely satisfying to me as a journalist and a citizen.

Individual landowners along the San Joaquin River have always been a force. When Friant Dam was being built in 1947, the bureau shut off the flows during the fall salmon run and that triggered the filing of the famous *Rank v. Krug* lawsuit. The Bureau of Reclamation was definitely the bad guy in the lawsuit. There was even a period of time when their attorneys refused to show up—and they were the defendants!

In the Ninth Circuit Court, Bud Rank and the other farmers won part of their suit and lost part of it, so the case went all the way to the Supreme Court. The Supreme Court issued a decision in 1963, but they sidestepped the real issues. It was finally decided that the Bureau of Reclamation, as part of the federal government, had a right to build a dam on the river, but that they should have negotiated with the landowners in advance of construction. That's what became of the *Rank* lawsuit, after all those years in the courts. Eventually, the bureau agreed to let enough water out to meet the needs of landowners down to Gravelly Ford, but the river dries up from there to Mendota.

In order to officially operate Friant Dam, the Bureau of Reclamation needed a permit from the California State Water Board, but they did not get permits until 1959. They operated the dam illegally for twelve years. During the Water Board proceedings the California Department of Fish and Game was emphatic about the fact that the San Joaquin had been a great fishery, not only for salmon but also for trout, huge sturgeon, and many other fish. The Water Board said, "We recognize that there's been great damage to the fishery but we think that the public interest is better served by diverting this water for agriculture." However, the board's specific language said, "We don't foreclose that someday a salmon run can be reestablished if conditions change or if the public attitude changes." The board authorized the bureau to legally operate Friant Dam, but they specifically left open the possibility that a fishery could be restored at a later time.

After the Water Board issued its opinion, the Department of Fish and Game wanted to go to court under the Fish and Game Code, Section 5937, a law that was passed in the 1930s to make sure that the fisheries weren't totally destroyed as all the big federal dam projects got underway. That code said that owners and operators of dams in California must let enough water through, over, or around the dam to sustain the fishery below in good condition. But that potential lawsuit was blocked by Governor Pat Brown, who was very beholden to the big agribusiness interests which dominated state politics for much of the last century.

We, the people of California, own our river water. That's written into our state constitution and our California Water Code, but in the 1930s and '40s we gave water permits to the Bureau of Reclamation because we desperately wanted the bureau to build dams. We gave the bureau water rights to seven million acre-feet of water. That's enough water for seventy million domestic users, and the Bureau of Reclamation does not pay the state one dime for that water. The Sierra snowpack contains several billion dollars in liquid gold. The state of California—which is currently strapped for cash and deep in debt—just gives that water away. The Bureau of Reclamation sells San Joaquin River water to corporate agribusiness, to Southern California developers, to subdivision developers, and to urban water districts. We need to understand that a lot of the San Joaquin River water is being used to build more hillside subdivisions in Southern California. Irrigation districts and even individual farmers can buy water for a fraction of its true market value and legally resell it to Southern California interests for ten to twenty times what they paid for it. It's windfall profits. The new cash crop in California is water.

Unfortunately, we are learning that our precipitation over the last century was an aberration. Believe it or not, the twentieth century was unusually *wet*. We have had epic droughts in the American West, which we are just now encountering again. Droughts can last for centuries. For the first two hundred years of this country's history, the American West was called "the Great American Desert." We have to acknowledge the truth of that. Sure, we have fourteen hundred dams in California, but what good can a dam do you if it sits empty?

The way we treat our rivers is a sign of how advanced we are as a civilization. A river is the lifeblood of any culture. Hydrologists know that if more than 30 percent of a river's water is taken, it becomes a dead river. The Hopi Indians have a saying about what is going on in America now, and it's a chilling thought. They say, "We are eating our grandchildren." I've never heard a better expression of what's going on in this valley.

Edmond R. Mosley, M.D.

Ed Mosley, a respected Fresno physician and community activist, settled in Fresno by chance and was drawn to fish in the San Joaquin River, where he still fishes after fifty-six years. He's a card-carrying member of the Fresno Hawghunters.

FISHING IS SOMETHING THAT I'VE ENJOYED since I was a very small kid, but I didn't fish here at all when I first got to Fresno. I was starting my medical practice and was very busy with that. I had been here for about two years when a friend of mine said, "Come on, take a break. Let's go fishing." So we went out to the San Joaquin River. I can't remember now exactly what part of the river we were on; I think it was down where the river comes close to the Municipal Golf Course. The river's not very big there and I wasn't accustomed to such a small river. The only rivers I knew were the Cumberland and the Wabash Rivers in Indiana where I learned to fish. Those rivers were not behemoths, but they were large. When I saw the San Joaquin, I said, "This is a river?"

I have to say that I came to Fresno by chance. I was planning to open a medical practice in San Jose and was driving there from Las Vegas. I got into Fresno late one afternoon and decided I'd just stay here overnight and finish the trip the next day. As I was here, I decided to look around, see what the city looked like and talk to a few people. One day led into two and two into four. I talked to the people at the medical society here and they encouraged me to open a practice in Fresno. They said, "We need you." I opened an office in West Fresno. I started off in family practice, but my training was as an internist, so as soon as my practice got big enough to support an internist I turned the rest of the practice over to other doctors.

Since that first time on the San Joaquin, I've had many very enjoyable times catching trout, catfish, and sometimes bass. I don't eat fish. I love to catch them, but I really don't like to eat them at all. My wife, Marion, loves fish but I throw back what I know she won't eat.

It's true that you can't always catch a fish. Anybody who tells you they have never been skunked while fishing has never fished much. Some days nothing is going to bite. Going fishing is like rolling dice; if you play long enough, you're going to crap out. I've used worms, mechanical bait, and lures. I will use anything that I think will entice a fish to bite. There have been plenty of times when I didn't get one bite, and there have been lots of times I got my limit and then some. When I say, "and then

"There must have been thirty or forty of us. We'd get together to fish and to tell fish stories." —Edmond Mosley. Courtesy of Edmond Mosley

some," it was when I was with somebody who didn't get any fish and I'd help him out, let him have a few of mine. Other than that, I'm careful to stay within the limits.

The easiest place to get access to the river is at Lost Lake. I can't get into all the other areas that I used to, climbing down cliffs and under the fences. I'm a little too old for some of those acrobatics. I can still get down the banks okay, but getting back up presents a problem.

I belong to a group called the Hawghunters. A hawg is what they call the biggest fish of a species. Let's say you're fishing for perch and you catch one that's about a pound—that's a hawg. You see, perch usually run only a half to three-quarters of a pound, so a pound is huge. If you catch a trout that's four or five pounds—that's a hawg. So we are the Hawghunters. We spelled it that way so nobody thought we were pig hunters. There must have been thirty or forty of us. We'd get together to fish and to tell fish stories. This was an organized group; we were even registered with the State Department of Fish and Game. We had to show evidence of our catch—our hawg. Unless we had witnesses, nobody believed us, and we took pictures too. Usually, if there were a bunch of us together, there was a prize for who got the biggest fish. It was an enjoyable group but I think the most fun I've had is with just two or three guys. Over the past twenty years I've had quite a few small groups of fishing buddies, but unfortunately they've all died. They died and left me here to fish by myself.

I don't think my stories are very interesting; they're just your basic fishing stories—stories fishermen tell fishermen. One night a friend of mine decided that we should just drift down the river in a canoe and see if we could get anything. We weren't expecting to get anything too large, so we had very light tackle. He hooked something and started trying to get it in. I'm sure it must have been a fish but we knew it was very, very large because it was pulling the canoe. We didn't want to break the line so we let it pull us for three or four miles. We finally got just opposite of Scout Island and my friend said, "I think I can get it in now." Just as he got it up to the boat, the line broke. It was dark that night so we never did see what it was. I imagine it was a big catfish; they're the biggest fish that I've seen in the San Joaquin. There are some very big fish out in that river.

"Anybody who tells you they have never been skunked while fishing has never fished much."—Edmond Mosley. Courtesy of Edmond Mosley

"I knew the creatures that lived here and I couldn't stand the thought of seeing them disappear." —*Clary Creagor. Courtesy of John Buada*

The Founding Mothers

Clary Creagor, Mary Savala, and Peg Smith are well known as the "Founding Mothers" of the San Joaquin River Parkway. In 1984 they were unknown to each other, but that changed when each concluded, in her own way, that river-bottom development on the San Joaquin River would devastate the riparian habitat, an environment hundreds of species of mammals, fish, and birds call home. The story of these three environmental activists, each motivated by her own particular view of the river, is one of perseverance and patience and passion.

Clary Creagor

We moved out to the river about thirty years ago, and honestly, I didn't even know there was a river here. We moved because we liked the view from our house—the mountains and the scenery. I didn't know anything about the San Joaquin River, but my son was quite small at that time and I would go exploring with him, making our way down to the river. We saw flocks of geese. There were thousands of geese then; they blackened the sky when they flew over. A little later on we discovered the heron rookery on Rank Island. We'd cross beaver dams that were so big and strong that you could just walk across them as though they were bridges. I had never been a bird-watcher until I had so many in my yard. There's a phenomenal number of birds and animals that come through the river bottom. I remember one Christmas that we observed forty-two mountain bluebirds in our yard during Christmas dinner. They weren't invited, but I was far more excited about them than our dinner guests.

There is still quite a diversity of wildlife out here, even now. This last year we had a gray fox family—three babies, a mom, and a dad—and they are the most entertaining creatures. The last time we were birding at Lost Lake, we saw *sixty* different species of birds. We've seen lazuli buntings on the river and the white-throated sparrow; we even saw a bittern, and they're very retiring, but this one was right out in the open. There is never a day that you can't see lots of birds. Bald eagles use the lake. There's a woodpecker tree. There's a huge flock of black-crowned night herons that park themselves in various places along here, along with the egrets and the great blue herons. You see them all the time, this stunning entertainment given to you free by nature. Over time, it's felt as though the river has been telling its secrets to me.

I guess it was 1984 when I read in the *Fresno Bee* that the Ball Ranch was going to be turned into a housing development, and it truly frightened me. By then the river was in my heart. I knew the creatures that lived here and I couldn't stand the thought of seeing them disappear. You can put houses anywhere, but you don't have a river everywhere. I felt that something had to be done and I thought, "What can I do? What can I do?" I called Chris Peterson, who was on the city council at that time, and he told me that Jan Ruhl, who's now Jan Mitchell, was also concerned about the river. I also called Dave Koehler, who was president of the Audubon Society at that time. Dave said he knew of somebody else who was concerned about development and its effect on the river. That is how I met Peg Smith, who had done a study on botanical species along the river in Lost Lake Park. Then I met Mary Savala. The three of us felt the same way about the river and we all felt that we had to do something to save it. I can't remember exactly how many meetings took place before we formed an official organization, but it couldn't have been more than a half a dozen. It all had to happen very quickly. There were events and deadlines coming up; everything was on a timeline, and we met constantly. That's the other strange thing about protecting natural resources; your entire time is spent in meetings, in offices, rather than at the river or anyplace in nature.

I quickly learned about twisting people's arms in a kind and gentle way. I'm not particularly brave in asking for something for myself, but if I was asking for the river, I was brave beyond what I ever expected of myself. Many wonderful people shared the same ideas and concerns and we drove each other on. I don't think any of us really knew how to go about doing this kind of thing, although we did have some good advisors. Hal Tokmakian, formerly of Cal State Fresno, advised us that the gravel companies were part of the economy of the county, and that if we opposed them we weren't going to serve ourselves well. Suggestions like that really helped us get started on the right foot.

I, who had never been president of anything, became president of the River Committee. I was known to be bossy but, other than that, I really hadn't a clue how to do that job. We came up with some solid guidelines and we agreed that we would always take the high road. That was important to all of us. There had already been a number of lies told about what we were trying to do and who we were trying to take advantage of. One of the first rumors that I remember was about the threat of condemnation of land. That was a major issue with the landowners. It was kind of amusing to us that people thought we had such power that we could condemn land. It was completely untrue.

Safety issues about Lost Lake Park were also a concern. The park had an unsavory reputation at that time. I've been coming to this park for thirty years, ever since we moved here, and I have to admit that when I first came to the park, it could sometimes be a scary place. The Hell's Angels came here a lot. But things gradually changed. The Southeast Asian families began coming here and they used the park as it was meant to be used, for picnics, fishing, and being together with their families in a place that's tranquil and beautiful. It became a safe place to come again; but the rumors persisted. There was speculation that the whole river was a dangerous, dirty place to be.

Landowners were afraid of trails crossing their land, in the mistaken belief that the Trust and the River Committee had powers way beyond their jurisdiction, that we could put trails wherever we wanted. We certainly had no such power, but landowners were afraid that people would trespass on their land and dump trash on their properties. Actually, such behavior existed long before the Parkway was established.

The other misapprehension was, "You'll never find the money to pay for it." We wondered about that too, but early on we gathered signatures for a state bond initiative for $5 million. The initiative passed, allowing us to purchase the Underdown property. That was our first big land purchase.

It was a gravel mine and now it has become a beautiful wetland. That success, before 1988, gave us some real credibility. Landowners were also afraid that the value of their land would be driven down by the Parkway, that they wouldn't be able to profit as much as they could if they could develop their land. We knew that we needed to protect the landowners' concerns and wishes, and we had to make them understand that we were not trying to take something away from them. We were actually going to pay them for their land. We pointed out that they really couldn't feasibly develop in the river bottom. Everyone remembers the terrible flood in January 1997, when the trailer park on the other side of the river turned into an island. Can you imagine what would have happened if the Ball Ranch had been developed with hundreds of homes? Most of the houses would have been completely underwater. The 1997 flood changed the way people thought about building on the river. People realized that development in a river bottom means that homes will be flooded—because that's what rivers do.

Once the River Committee was formed as an official organization, we held public meetings to discuss relevant issues for the river bottom: flooding, sand and gravel extraction, diversity of species in the river bottom, and reclamation. We hadn't even brought up the idea of a Parkway yet, but a lot of other things were happening at the same time. Don Furman came in as our executive director, and he immediately started taking people out on canoe trips. Those canoe trips got people out on the river and that, to me, was the very best way to sell the idea of a Parkway. There was very little public access at that time, but the river itself belongs to the public, and to actually be able to travel on the river helped people understand what the issues were. We all worked very hard to offer as many people as possible a chance to experience the river firsthand.

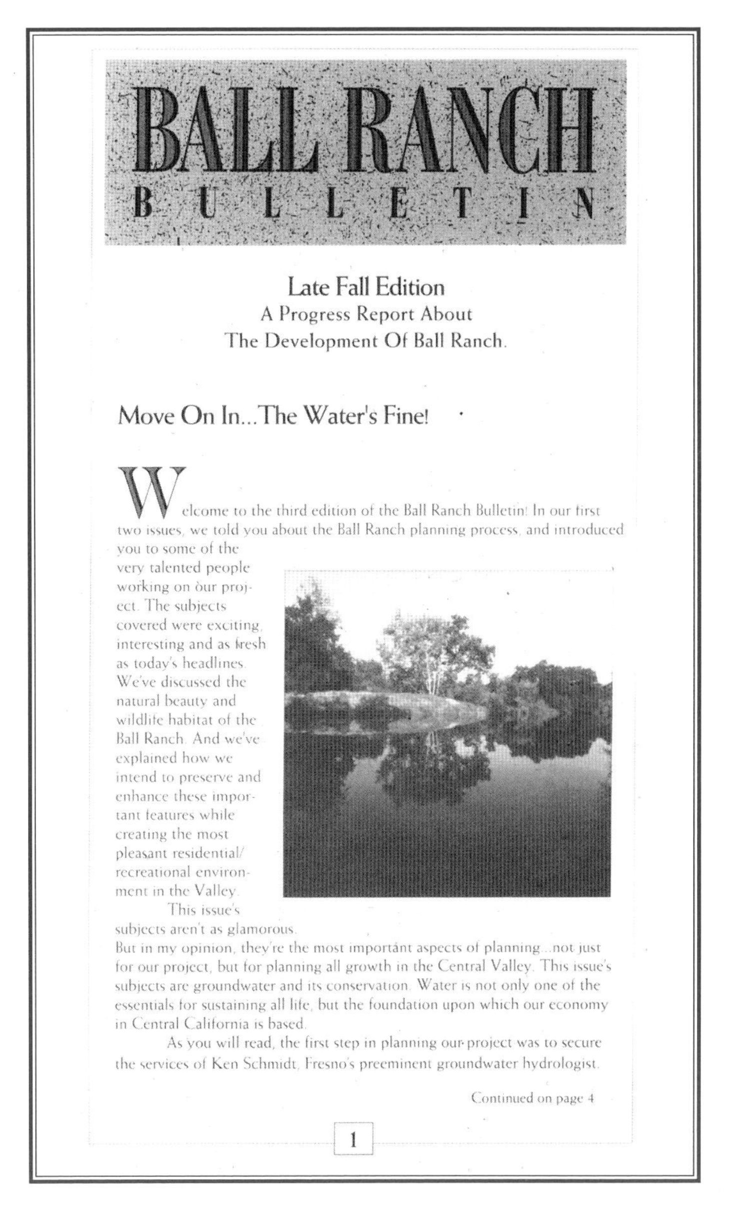

BALL RANCH
BULLETIN

Late Fall Edition
A Progress Report About
The Development Of Ball Ranch.

Move On In...The Water's Fine!

Welcome to the third edition of the Ball Ranch Bulletin! In our first two issues, we told you about the Ball Ranch planning process, and introduced you to some of the very talented people working on our project. The subjects covered were exciting, interesting and as fresh as today's headlines. We've discussed the natural beauty and wildlife habitat of the Ball Ranch. And we've explained how we intend to preserve and enhance these important features while creating the most pleasant residential/recreational environment in the Valley.

This issue's subjects aren't as glamorous. But in my opinion, they're the most important aspects of planning...not just for our project, but for planning all growth in the Central Valley. This issue's subjects are groundwater and its conservation. Water is not only one of the essentials for sustaining all life, but the foundation upon which our economy in Central California is based.

As you will read, the first step in planning our project was to secure the services of Ken Schmidt, Fresno's preeminent groundwater hydrologist.

Continued on page 4

1

Marketing brochure for Ball Ranch development. Courtesy of San Joaquin River Parkway

The Ball Ranch development created a lot of controversy in the community. Many people sided with the Ball family and I understood their concerns, but it was a battle we absolutely had to win. We had to garner the public's concern for the river, clarify what was at stake, and turn out people for public meetings. We lobbied elected officials, for we had to prove the value of the river to them, too. We went to the planning commission and the board of supervisors twice on this issue. One of the meetings lasted for something like eight hours; it had to be moved to a bigger room. We had broad community support; we even had Girl Scouts up there testifying.

We realized that we needed a parkway to protect the river and that we needed to fight for something positive, not just against development. We went to a meeting of the Land Trust Alliance. At that time there were already something like six hundred parkways across the country. We didn't have to reinvent the wheel, thank goodness. By then we had the help of Assemblyman Jim Costa. We also had immeasurable help from the *Fresno Bee*. Reporter Gene Rose, who later served on the board of the River Parkway Trust, wrote a series on parkways across the country for the *Bee*. I think these articles really had an impact on the public. They say a picture is worth a thousand words, and those articles were full of pictures as well as words. You could look at the photos of the river and think, "Maybe I need to read this and find out more."

It's been a great lesson in democracy. I had always voted but I didn't really know who my supervisors were; I didn't know about local politics. Now I care. The quality of daily life is dictated to a great extent by the choices local officials make. It is in everybody's interest to make sure that these officials support all of our interests. I really do believe that one person, or a couple or three people, can make all the difference in the world if they are resourceful and clear about what it is they want to do.

I think the public, including landowners on the river, is very happy about the Parkway's progress. By now, most of the land that's been purchased in the river bottom was purchased with public and private funds and it was bought at fair market value. The Parkway has been responsible for bringing this funding to our valley and for turning the river over to its public.

People often used to say to me, "Well, if you don't believe in development of the river bottom, why do you live there?" I always thought that was a reasonable question to ask, but then I thought, "It's damned lucky that I live in the river bottom, because I understand what the river means to the quality of life here and why it is important to protect it."

What now? Well, we need to get more people down the river in a canoe. It's pretty hard to turn away from a heron rookery, or even a beaver dam. The river environment exists simultaneous to the lives of those who came before us and those who will come after us. Each generation has the opportunity to understand the river and to form their own experience.

Mary Savala

MY HUSBAND AND I ACQUIRED our property on the river bluff about forty years ago. The choice was kind of haphazard; the place just happened to be available but I was delighted because it seemed like a very special location. At that time there were plans to have a forebay for Friant Dam below our property. Our vision was that we'd be half a block away from this wonderful lake; I could picture the little boats and our four kids having fun in the water. Of course, that never happened; and that's a good thing, I can see that now. But we had the river and we spent a lot of time down there. All the land from our house to the river is privately owned, so we had to get permission from our neighbors to slide down the bluff.

The part of the river that lies directly below us is the Mullin property, and people have always used that passage with their permission. It was really pretty arduous getting down there, but my kids could do it and that was forty years ago, so I was still pretty limber.

I remember one time I was working on my income taxes and I was sick and tired of being inside. I decided to go for a walk by myself down to the river. I scrambled down the bluff. It had recently been very wet, and although that day had cleared up and the sun was shining, it was still pretty muddy. I was slipping and sliding, climbing back up the bluff, and I fell. I was lying there spread-eagled on the side of the bluff thinking, "Nobody knows where I am and I'm really having trouble here." That desperate thought prompted me to work a lot harder and I did get home all right that day, but getting access to the river has always been a problem for the whole community.

A friend of mine who worked in the city planning department, Jan Ruhl Mitchell, called me one day and said, "Did you know that we have a proposal to put in a number of homes and businesses, including a hotel, down on the Spano Ranch? How do you and your neighbors feel about that, Mary?" My first reaction was, "Well, it's probably inevitable." Then I got this terrible feeling. I started realizing that telephone poles, roads, and sewer and water pipes would have to be built in the flood plain. Dogs and cats would be introduced into an environment with so many native birds and animals that need a peaceful life. It didn't sound like a very good idea at all.

Jan put Clary Creagor and me together, and soon Peggy Smith appeared on the scene. I had been moved by the Spano project but Clary and Peggy were primarily motivated by the plans to develop the Ball Ranch. It was all happening at the same time. Fresno was expanding north and northeast, and the developers were anticipating that. Those two large projects came along in the same year. We all sat down in Peggy's living room saying, "What do we do?" Right away, it became obvious that a small handful of people couldn't just oppose development piece by piece. We'd never get anyplace that way. We weren't sure what our role could be and we found it all pretty overwhelming. We brought Hal Tokmakian in, who was a former planning director for Fresno County, and we asked him, "What do we do?"

It was Hal who reminded us that there was a city plan for Fresno that had been drawn up in 1908 by a man named Henry Cheney. He'd been asked by the city of Fresno to do a plan and, among other things, he'd mentioned that people were using the river for recreation, even then at the turn of the century. People would come down to the river to cool off, to play. There were commercial recreation sites and a trolley that took people out to the river. Henry Cheney talked about what a wonderful place the river would be for a city park, and he recommended that there should be a plan established to do that. We decided, "Okay, let's be *for* something!" Let's create a proposal for a park, something that we can really get behind.

That's how the idea of the Parkway really began, but we had big obstacles. One was acquiring land. Property owners were very skeptical of how their property rights would be affected by this, and so we had to make friends among those people. We also had to galvanize a community. A lot of people, especially newcomers to Fresno, didn't even know the river was there. We had to start talking about the river.

Chuck Peck came up with the idea of taking people out on canoe trips; Don Furman came into our lives at that time and helped us organize those. He knew a lot about canoeing and guiding people down the river. We'd have pancake breakfasts in the morning and then we'd canoe for hours. Everybody

"Why would we build on the river? In my way of thinking, it's like burning your antique mahogany table because you're too lazy to go out to the garage and get firewood out of the woodpile." —*Mary Savala. Courtesy of San Joaquin River Parkway*

loved it. It was the hook for so many people. This river was so close to a busy city and yet so serene, so peaceful. We made lots of new friends and we made money too, to help us pay for our printing and mailings. We had $1,800 from the very beginning, from people just randomly sending us checks, and I remember looking at the money and thinking, "What are we going to do with this?" We had to get organized and be responsible. We formed an official committee, brought in other people, created bylaws and articles of incorporation.

Finally the job really got to be too big for us. At that point we had the wisdom to hire an employee to direct our efforts. Don Furman had political experience and this was at the time when we had just begun to get political, lobbying the board of supervisors and the city council. Don liked politics and he had a lot of history with the river and the community. We agonized about how to pay him, but he agreed to accept the job under shaky circumstances. That was another great leap forward.

We wanted to acquire land, but who was going to own the land? That became the next issue. Peggy and I went to the National Land Trust Alliance at Asilomar State Park in Pacific Grove. We had our eyes opened to the fact that there were six hundred land trusts all across the United States who held land in trust: a one-acre hillside, a garden for growing vegetables, a huge tract of land in Montana, a narrow section around the Chesapeake Bay. We could see how this idea might work for us. We talked to the Trust for Public Land (TPL) and they patted us on the hand and said, "You've got a lot of work to do, ladies."

With TPL's advice, Don Furman put together a board of directors. We asked Coke Hallowell to join us, and Coke agreed to be our president. We had two bankers and some people familiar with real estate. We had a very diverse group of interests. Our passion to save the river was our uniting thread. Of course, you want talented people involved in doing the work, but most of all you want *passion*. I remember that first meeting of our new trust; we all got together to sign the articles of incorporation in Rebecca Gomes's backyard. I looked around and didn't know anyone. I remember thinking, "Who are these people and what is it they want?" Within six months we were a good working group, trusting each other, making phone calls and raising money. That was another big leap forward. Of course, now the River Parkway Trust has become a well-respected, credible organization, and the whole effort to save the river has become fashionable and accepted.

I have to mention Congressman Jim Costa, how valuable he was in championing our cause and raising the millions of dollars it's taken to acquire land. I see Jim not as a politician but as a statesman. It was Jim who suggested forming a conservancy. At that time there was the Sierra Conservancy, the Coastal Conservancy, and there were conservancies in Santa Monica and Tahoe. These were state agencies that managed property in the public interest. The San Joaquin River Conservancy was formed and the guidelines called for two public members. I served as one of those public members.

During all this, I wound up on a few enemy lists, but I have a wonderfully supportive husband who would tell me, "Stop feeling sorry for yourself." He'd urge me to stretch myself. I remember standing up in public meetings with my stomach churning, my voice shaking, and my knees buckling. Rudy would tell me, "Just imagine all those guys naked." That would sure take away the formality. There were a couple of people out there with big pot bellies and that would just really make me smile. I remember there was a person on the policy board who used to make me tremble; when I'd go up to speak he would say, "All you people," when I'd be standing there all by myself. One day I finally had the courage to turn around, look behind me and shrug my shoulders. The whole room laughed because it was obvious that it wasn't *all you people,* it was just me.

There are still some things that we have to deal with. The big challenge for us now is how we will operate and maintain all this property. There will always be people who want to put football fields down at Lost Lake Park, and we'll have to keep talking about the guiding principles for the Parkway. I have always envisioned it as a passive, natural place where people could come to birdwatch and fish.

We could be so much smarter about where and how we develop. We have this wonderful waterscape and riparian resource, and we still have lots of land left to build on. Why would we build on the river? In my way of thinking, it's like burning your antique mahogany table because you're too lazy to go out to the garage and get firewood out of the woodpile. It would be a horrible waste. This river is like a miracle. All we have to do is take care of it. My mother said, "If everybody would do just a little bit, then the world would be a whole lot better and nobody would have to do the whole thing." So I think this can be my little bit.

Peg Smith

I WOULD LIKE TO BEGIN WAY BACK, in Los Angeles, where I grew up. We were not too far from the Los Angeles River and I remember, as a child, looking at that poor pathetic river and those enormous concrete banks. Apparently the concrete was put in after a major flood in 1938 or '39. A lot of homes had been destroyed and possibly even a few people had died. Actually, the concrete was pretty handy for a kid who liked to ride a bike; I used to go bike riding on the banks near Griffith Park. But the fate of the Los Angeles River broke my heart. Even as a child, I could see how terrible that was.

Many years later, when I heard that people were seriously considering putting houses in the San Joaquin River bottom, I thought, "That's what will happen. They'll put in the houses, there'll be a flood, and then the whole river bottom will be concrete." That image was what really motivated me at the time.

In the 1970s, when my children were small, we went out to Lost Lake Park quite a bit and the kids played in the river. I was active in the Fresno Audubon Society, which had built a nature trail that we spent a lot of time exploring. While taking some biology classes at Fresno State, my friend Joyce Hall and I made an effort to do a complete survey of the plants of Lost Lake Park. We went every week for a year and noted all the native plants, not the introduced plantings or the lawns or the exotic shrubs. I think we ended up cataloging something like 175 different species.

So I had strong feelings for that part of the river and was just totally blown away and horrified when I found out that people were proposing to put an urban neighborhood on that land. I read the little article in the *Fresno Bee*—when I say *little*, I mean it was on page B9 or something, just barely a mention. A representative from the Ball family had spoken to the board of supervisors and asked for pre-approval before any official activity took place. Fortunately, the *Bee* had a reporter there; he wrote about the request and that was the beginning of my activism. I was in a spot to make a difference.

At first I was in shock. I thought, "Somebody needs to do something." The only thing I could think to do at that moment was to call the president of Fresno Audubon, who happened to be Dave Koehler. I said, "Dave, have you read about this suggestion that the Ball Ranch be paved over and filled with houses?

What's Audubon going to do about that?" Dave said, "Well, I don't know much about it, but the funny thing is I just got a phone call from another Audubon member. Do you know Clary Creagor?" I didn't know Clary, but Dave gave me her phone number and we agreed that we would meet in a couple of days in the coffee shop of the Del Webb building downtown. At that time I was a paralegal working for the county and it was easy for me to duck out of the office. I met with Clary Creagor. She had already met with Mary Savala, and I believe Mary had already met with Hal Tokmakian. As the years went by, we all became very close, but we started out as complete strangers.

In about 1987, we decided that we really needed to get the word out in an organized fashion. Hal Tokmakian was involved in city planning and he was very important in helping us understand the nuts and bolts of what we needed to do. I can't give him enough credit for that. Mary was the expert on the governmental aspect of the whole thing; she understood processes that I did not. Clary was the fighter; she was really in there, going to get 'em. She was willing to be president of the River Committee and speak before lots of people. I was definitely not the person to do that, but I could provide a newsletter. It was six pages long, printed on blue paper, 8 ½ by 11, simple. We mailed them to the entire Audubon mailing list and to many other people as they showed interest. I do take a certain pride in that. That was my biggest contribution and it was good.

The newsletter existed for about five years, right into the time that the Parkway Trust was officially formed. We announced fundraising activities and speakers and published articles about the biology of the river and what made it so special—the plants and animals and birds. Perhaps most important were the articles about the progress of the Ball Ranch project and our efforts to stop it. We'd announce when

• ALERT • ALERT • ALERT • ALERT • ALERT • ALERT • ALERT • ALERT • ALERT • ALERT •

ALERT!

On Tuesday, July 27, at 2:00 p.m.,the Fresno County Board of Supervisors will consider the final approval for the Ball Ranch project in the San Joaquin River bottom. The project's developer, the Sienna Corporation, has submitted a final version that fails to provide several of the special conditions which the Board of Supervisors called for when they approved the project. These special conditions include provision for a point of public access to the river and a firm commitment for the construction and maintenance of a public trail through the project.

Please attend this hearing! The Sienna Corporation must be required to honor the conditions for this project!

If you cannot attend the hearing, please phone the Board of Supervisors (488-3531) and express your concern for Sienna's obligations in the river bottom.

July 27th • 2:00 P.M. • 3rd Floor, Hall of Records

• ALERT • ALERT • ALERT • ALERT • ALERT • ALERT • ALERT • ALERT • ALERT • ALERT •

Rallying support, these cards were enclosed in the monthly San Joaquin River newsletter. Courtesy of Clary Creagor

there was going to be a hearing and try to convince people to attend. In at least one newsletter, we enclosed a postcard for people to express their opposition to their representative on the board of supervisors.

The first developer who wanted to build a large development on the Ball Ranch was a man named Danny Brose. He'd come from Palm Springs and, I guess, had been successful as a developer there. He seemed to think that everything here was pretty much like it was in Palm Springs, except that there was a little trickle of a river someplace. He offered a tour to several members of the River Committee's board of directors. We went in a little bus and I vividly remember his pointing out that he'd "tear out all those ugly shrubs over there." We were all saying, "No, no, no! The animals and birds need those shrubs!" This was the developer that had been hired by the Ball family to build their development. They did an environmental impact report and the scientific part was absolutely awful. The River Committee was very active in responding when that came out. In fact, we produced dozens of pages of comments on all the things that were wrong with Mr. Brose's EIR. After that, he disappeared. And for a year and a half or so we actually thought it was possible that the project wouldn't be revived. But no. After a certain period of time it came back in much greater force because by this time the Sienna Corporation had been hired to do the project. They were a much larger corporation—more money and expertise—and a much more formidable enemy. They stayed around for a very long time.

Our group was writing letters expressing our point of view, but there were plenty of letters on the other side as well, saying that this development was going to be an asset. There was a strong element of support for the Balls. They were an old-time ranching family, so there was sympathy for them and for the whole idea of doing whatever you like with your own property. But by this time the Balls had already permitted their land to be dredged for gravel and they had made a lot of money on that. So I didn't personally worry a lot about that aspect of it. In the end, I think the Balls did pretty well.

It was probably 1986 when Mary, Clary, and I went to a rally of the Land Trust Alliance that was held at Asilomar State Park. That's where we learned about land trusts. That certainly was the beginning of the process of setting up the San Joaquin River Parkway and Conservation Trust. It was really all about saving the land; these wonderful wild areas like the messy, jungly, shrubby Ball Ranch would remain the way they were, and the wildlife would continue to have a place to live. About that time I happened to fly over the river and I was struck with how that ribbon of green came down through all this dry brownness—the Fresno plain. I did not want it to be covered with rooftops when one looked down from above. I was inspired to keep it green.

The day that I remember most vividly was the board of supervisors' meeting when they finally were ready to vote yea or nay on the Ball Ranch project. Through one of our board members, the River Committee had set up a telephone bank at a stock brokerage firm, and on the night of the hearing we called hundreds of people. The crowd we got to come was so large we couldn't believe it. The *Fresno Bee* estimated four hundred people attended the hearing. One of the other things we accomplished, without a doubt, was to show the leaders of Fresno City and County that the river was there and that it was an incredible community asset.

Many years later, Assemblyman Jim Costa held monthly hearings that involved many different river-bottom-related agencies. I attended as the representative of the River Committee. Somebody showed a publicity movie about a development idea for the Spano Ranch. It was the most amazing propaganda I had ever seen. It was a presentation about a beautiful English village on the shores of the Thames River, how lovely the houses were and how the Spano Ranch project would be just like the Thames River Valley in England—a huge development with hotels and things. I couldn't believe it. It showed such total disregard for the geological and environmental facts. I thought it was unbelievably ignorant.

I've sometimes asked myself, "What would have happened if the Ball family had *not* decided to put a development on their land?" Something else would have come up, that's for sure, but the Ball Ranch proposal definitely energized us. That's what got the whole Parkway effort going. I remember telling a friend at that time, "If we can get to the point where the idea of building houses in the river bottom is absolutely unthinkable to this community, then I'll feel that we have accomplished what we need to accomplish." The work continues.

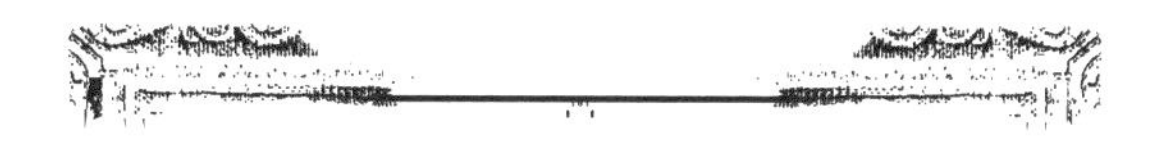

Hal Tokmakian

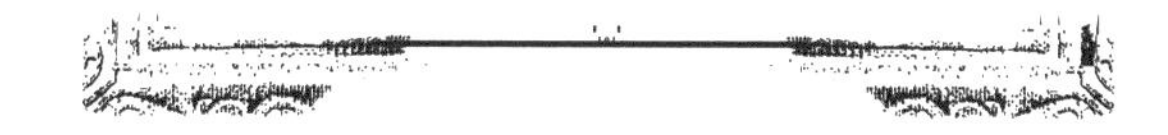

At one point, early in their efforts to protect the river, the Founding Mothers called upon Hal Tokmakian to help them understand the details of county planning issues and environmental impact reports and how to respond effectively. Hal has served on the Fresno County Planning Commission and has taught university classes in regional and urban planning. He is involved now in many civic activities, always contributing greatly with his vast knowledge of responsible planning.

WHEN I WAS A KID WE'D DRIVE UP TO the edge of Van Ness and down into the river bottom. There was a park there; I don't know if it was a formal park, but there was at least some access to the river. I remember Lane's Bridge too; we'd cross it on our way up to Bass Lake. That's about all I remember from my early days at the river. Fast-forward a little and we come to my professional life as a city planner. I have all kinds of river memories from that time.

I returned to Fresno in 1958 after college. I got a job as a senior planner with the county planning department. Those were some interesting times that to some extent determined the future of the San Joaquin River bottom. One of the first projects that I was responsible for was bringing Fresno County's Park and Recreation Plan up to date. Part of my job was to prepare a reconnaissance study of all the resources that were suitable for recreation and park purposes, from one end of the county to the other. We photographed the San Joaquin River from the town of Friant all the way down to north of Kerman and identified the resources and points of access. Also in 1958, there was an unprecedented joint undertaking, a planning study that was a cooperation between the city of Clovis, the city of Fresno, and Fresno County. That plan also identified the San Joaquin River as a valuable resource. The county was not spending an awful lot of money in establishing and maintaining county parks. Nevertheless, our plan was adopted, and

"It's not difficult to take a stand when you have a great resource staring you in the face."—Hal Tokmakian. Courtesy of Fresno County Library

1958 was the first year that the San Joaquin River bottom was officially determined to be an important resource in Fresno County.

I emphasize the phrase "river bottom" because we were not just concerned with the riparian habitat along the river and its edges; we were equally concerned with the bluff-to-bluff conditions. We were trying to balance the three elements: the important recreation and riparian habitat, the concerns of agriculture, and the sand and gravel resources. Gravel mining had become quite controversial because there were many use permits being granted along the river. We knew that one day the mines would be exhausted. If the county and city of Fresno administered the planning laws properly, this land should be reclaimed for beneficial use to the public. Believe it or not, in the final Fresno-Clovis Metropolitan Area Plan, the whole river bottom was painted green on the map. There was a strip of green shown all the way from the Fresno metropolitan area to Friant, and the planning designations identified the area as an open space, multi-use resource.

Not much changed along the river from 1968 through the seventies, but during the boom days of the early 1980s, big construction projects were happening all over the state, and in Fresno County land speculators were out looking for development opportunities. That triggered a crisis in the minds of many conservationists. The Ball Ranch project was proposed along Little Dry Creek as it enters the San Joaquin. A lot of people became concerned about this development. We all knew that if it was allowed to happen, it would get right to the heart of an important open space resource. It was potentially very, very profitable to convert the ranch into a golf course and estate homes, and money is a powerful force to reckon with.

One day I got a telephone call from Mary Savala. She was active in the League of Women Voters and knew my wife, Barbara, who was president of the league at the time. Mary was very concerned about the Ball Ranch development plans. She said, "We've got to do something. What do we do?" I agreed that this was a precedent-setting issue. This would break the back of anything that we had planned for the public benefit in the San Joaquin River bottom.

By then I was a professor of city and regional planning, and since I was a former planning director, I suppose people thought that I had something to contribute, that I could help deal with the plodding pace of the planning and zoning administration. The Ball Ranch was in an unincorporated area, under the jurisdiction of Fresno County. I knew the issues that had already been studied years before and the many steps that still had to be taken. Fortunately, by that time the state of California had passed the California Environmental Quality Act (CEQA). The act, adopted in 1970, required environmental reviews of private development projects. The Ball Ranch project was undertaken by an outside corporation that saw grand opportunities here, but we knew that they would have to take many steps along the way, including the administration and processing of a CEQA report and the preparation of their environmental document. We knew that in the process we would have an opportunity, as a public citizens' group, to get into it and truly participate, review, comment, stand up and have ourselves heard, which we did.

We began to review and bird-dog the plans that were being created by the applicants for the Ball Ranch project, making sure that we had all the information to properly understand what was being done and how it was being done, what the next steps would be, what the timetable looked like. Since I have planning experience and none of the others had, I got seriously involved.

Things were beginning to get stirred up and the controversies were starting to put pressure on the politicians. The Ball Ranch process went on and on and on, and eventually the board of supervisors, even after a lot of strenuous opposition from interest groups, approved the developers' plan. They had all the necessary permits and were working through the final subdivision maps, and there it sat. But the controversy didn't resolve itself; the environmental documents that the developer submitted were appealed, and then, fortunately, the financial conditions of building and development in the United States took a downturn.

Eventually the rosy years of the 1980s faded away and the markets troughed. Nature helped us along the way, too; we had a flood along the Little Dry Creek and that changed attitudes a little. There was a year of heavy rain when the San Joaquin River bottom flooded, and that gave a new outlook to the risk involved in building on that particular project site. Conditions changed, attitudes and restrictions changed, and the developers finally backed off. A lot of the Ball Ranch has since been acquired by the River Parkway Trust and the state of California, so it's out of the running for a big development—definitely a good thing for the people of this community. There's now a huge amount of public support to preserve the area around the river but we have to be careful not to become complacent.

I've learned that we can't depend on our elected officials to hold the line. It's a game they continue to play—the zoning game. I believe that there are no greenbelts without greenbacks. That means that property rights acquisition, in one form or another, whether it's an easement, a trust, or an outright acquisition, is the only way to ensure that open space will be retained. Over the years, that's become increasingly true; that's the way it is today all over the state of California. That's the new ball game.

There really was some amazing foresight in 1958; those original plans provided the background information for those of us who, thirty years later, went on to fight development in the river bottom. Even back then you would be surprised by how easy it was. It's not difficult to take a stand when you have a great resource staring you in the face—you just take your green pencil and color it in, bluff to bluff.

San Joaquin River

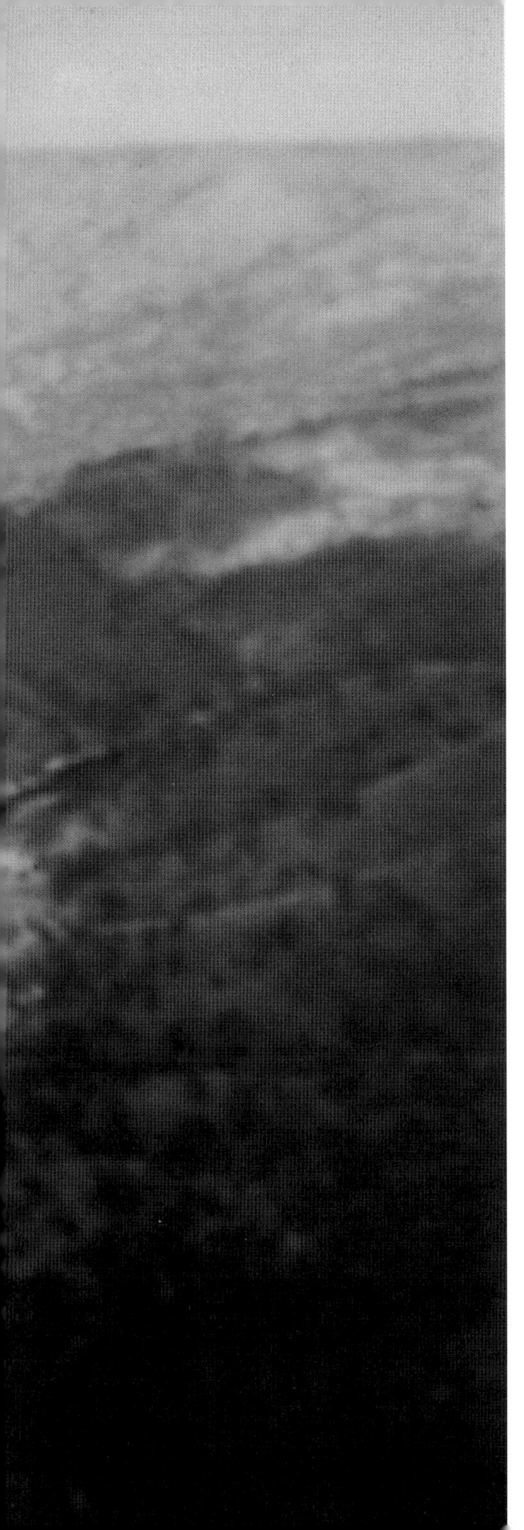

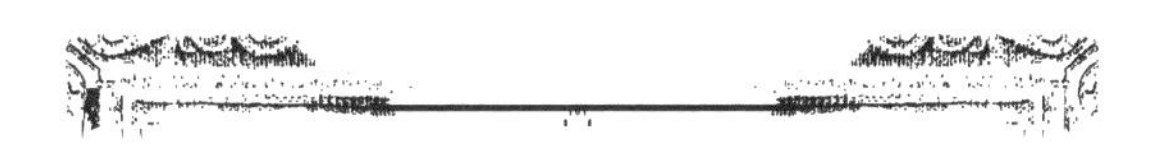

Paula Landis

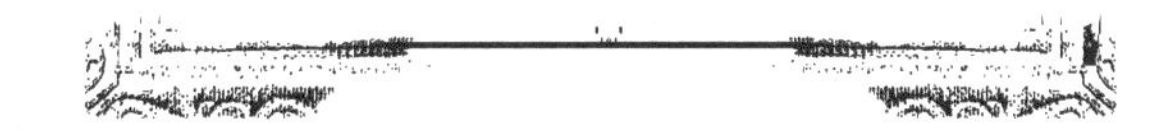

Paula Landis is chief of the San Joaquin District for the California Department of Water Resources. This key role, along with her personal history with and love for the San Joaquin River, puts her in the perfect position to help elucidate the next stages of the river's restoration process.

MY EARLIEST MEMORIES GO BACK TO visiting my godfather, Dr. Earl Lion, out on the river bluffs. He was chairman of the English department at California State University, Fresno. He hired many of the people who made great contributions to the English department, and my father was among them. Earl never married and never had children, so all of us faculty children became his little godchildren.

In the early 1950s, Earl and James McClatchy planned a development out on the bluffs, on the Fresno side of the San Joaquin River. I think they did a wonderful job planning it. They had arranged the houses so that nobody would block anybody else's view of the river and their plans included other very forward-thinking ideas. Earl built the first house out there, a beautiful adobe home with lots of windows. I know he struggled with that whole process; when he'd see a big piece of heavy equipment, he said he'd get a knot in his stomach. They had spent so much money getting it all together, planning and putting in the streets, and then nobody bought out there. They couldn't even sell the lots for $2,000. That just shows you how much things have changed since then.

But I loved it out there. Seven or eight of us kids would ride our bicycles and explore the whole area. It was all open, not a house around. We would climb down from Earl's property, which was on the bluff, straight down to the river. It was

"Rivers are dynamic; the water flows and so does everything else." — Paula Landis. Courtesy of Eastern Fresno County Historical Society

hard to get out there. I remember getting scraped up by the blackberries. At the bottom of his property there was rock, a gravel bar, and we'd hang out there for hours. We found great swimming holes and we attempted many times to canoe the river. We'd try to go down the stream but we'd run out of flow, get stuck, and have to drag the canoes back up. One thing that has always stayed with me is the smell of the river. I love that scent. Whenever I approach the river, when I get to a spot at a certain elevation, it just hits me. You feel the coolness; then you get a whiff of that aroma. I guess it's kind of anaerobic; some people are not as receptive to it, but I love it.

While I was studying engineering, I became interested in water. I had a student position at the Department of Water Resources. I initially worked on agricultural drainage, which is all about stagnant, polluting water—very different from streams. Then Bill Loudermilk from the Department of Fish and Game came around looking for someone in the valley who knew about rivers. I got myself trained and I did my first restoration in 1989.

That was also the year AB 3603, signed by Assemblyman Jim Costa, established the San Joaquin River Management Program, or SJRMP. We were asked to look at the San Joaquin River from Friant Dam to the Delta, including the major tributaries up to the first dam in the High Sierra. It was supposed to be a consensus-based process involving a number of government and local agencies: agricultural and environmental groups, public agencies, flight control districts, park districts, and cities and counties. It covered the entire reach of the river and didn't have the political boundaries that a lot of other organizations are limited by.

We had an assignment to develop a plan in five years. In that time we came up with eighty actions that would benefit the San Joaquin River. It took a lot of talking to agree on language so it could be consensus-based. Then we spent a lot of time trying to prioritize things, but we finally had to give up on that. The actions are listed alphabetically in the plan.

Any action on the river is going to have impact downstream and probably upstream, too. Rivers are dynamic; the water flows and so does everything else. There are so many issues to consider: water transfers, the sale of water, water quality standards. There are the westside drainage issues and the eastside cities with their wastewater treatment plants and diversions. At the Department of Water Resources and the Bureau of Reclamation, we pump water out of the Delta; the river's flow significantly impacts our ability to pump. The whole state is affected by this river; it always has been, but people are just now finally realizing the significance of the San Joaquin.

There's a misconception that if you just add water, you have restored the stream. It's much more involved than that. The focus of restoration on the San Joaquin has been on anadromous fish because that's the law; it's the hook for restoration. Anadromous fish are those that live their adult life in the ocean, but are born and die in fresh-water streams. That's not limited to salmon; there are also sturgeon and steelhead and other fish that do that kind of migration. But we have to have the whole system: the vegetation that will shade the water to keep it cool, the diversity that provides habitats for the different life stages, and channels for the fish to get up and down the stream, past the obstacles, whether they're adult or smolt. In any system, every part has a role. We may see floods as very devastating to people and property, but even floods have a purpose. They scour older vegetation and plant seeds to start new vegetation. You have the age succession, which is necessary because certain species like old vegetation while certain species like new. The gravel needs to be cleaned. It's all part of the natural process.

Restoration of the San Joaquin is going to take a lot of commitment, not just on the part of the farmers and the environmental community, but also from the people of the state of California. Everyone in this state is connected to the San Joaquin and we have to help taxpayers understand that they are beneficiaries of this restoration. I think the valley understands that, as does the Bay Area, but I don't think Los Angeles sees that yet. I think the average person in Southern California sees no connection between himself and the San Joaquin River. People in Los Angeles don't think about where the water comes from when they turn on their faucets. We need to educate people.

I didn't understand an inkling about California or the valley or Fresno until I got involved in water. It was a real eye-opener when I learned how the state's river system was plumbed and how important water was to California. I've been in this business for eighteen years now and I still wonder and worry about the people who have to make decisions about water. I've focused on the subject, but even I can see that it's very tough to make a good call. There are decision makers out there who have to make very important determinations about our present and future water needs, and they only have a few years to learn a little bit about the issues. I don't envy them that responsibility.

John Buada

John Buada has spent many years honing his diplomatic skills while negotiating with private landowners, government entities, and gravel-mining companies. While spending years along the San Joaquin River, representing many different companies, he has led children and adults on nature walks and photographed the abundant wildlife found in these rich habitats.

I STARTED WORKING FOR THE Fresno County Planning Department in the mid-1970s. They were working on their General Plan, and the first portion of that process was to create a plan for open space. They wanted to do specific regional plans for each of the rivers: the Kings and the San Joaquin. I was assigned to write the entire 1976 General Plan. I was fresh out of college, a San Francisco kid who had just come down here, and it was all brand new to me. I naively went out with the thought "A river is a river." I had a lot to learn. I started talking to people and found out that the attitudes of residents on both rivers were very different. At that time, people on the San Joaquin River were a lot more open to potential open space uses. They were in tune with the urban environment, while the people on the Kings River said, "This is our river. We moved out here to be away from people. Go away and don't tell anybody we're here."

At that time, the minimum parcel size was 100,000 square feet, which is 2.3 acres. Consequently, little rural subdivisions were popping up all over the place, which is not a good use of land at all. Residents had put in their own septic tanks and wells—totally inefficient. One of the things that we did in our planning was rezone all of the river bottom and agricultural land to twenty acres. That was not very popular at all with a lot of rural landowners. We conducted many public hearings and there were a lot of angry people. I remember once when my wife brought our two very young daughters down to one of the hearings. They were sitting in the back of the room and after the meeting I said, "Well, how did things go?" And my wife said, "That was scary. There were people that were so mad at you they were saying they wanted to take you out and hang you." My daughters heard that. That was tough.

It's interesting how ideas evolve over time. When the county updated the plan around 2000, people ridiculed the zoning restrictions. They said, "There's only a twenty-acre limit? It should be more than that." Twenty-five years after the General Plan was first drawn up, many people are saying,

"The article told a story about how the albino deer were special to Native Americans, that they were considered spiritual."—John Buada. Courtesy of John Buada

"Why didn't you make the limit higher? You did a terrible job on that General Plan."

Little did I know, when I was working on that open space plan, that this river was going to define the rest of my career. I left my job for the county in 1979 and started my own company. My first client was Don Underdown and his son Greg. Theirs was the first sand- and gravel-mining site that I worked on, and I helped them get the expansion for their permit in 1979. When I first went down to see what they were doing, I thought, "What is all this? They're digging holes everywhere." One of the things that immediately caught my attention was that there was still an amazing amount of wildlife around, in spite of the mining equipment, lots of excavation, and big trucks driving through constantly. I was really intrigued by that. How could these two disparate worlds possibly coexist?

One of the primary things that I do for companies is design reclamation plans. Our goal is to balance out the recovery of a natural resource—in this case gravel—with the need to preserve the other resources out there, namely wildlife habitat, the groundwater, and the air. We all use gravel and nobody thinks about it. In Fresno alone, we use five million tons of gravel every year. So we're trying to figure out how you can extract that resource and still protect the other resources. That's my job.

You mine the property, and then you restore or reclaim it in a continually evolving process. We try to reshape the land in a natural way and then we put in native plants and elements that will attract wildlife. Any changes we make to the land are done to improve it and to restore it to an even more natural state. We're constantly working on how to efficiently make it look natural.

We finished mining at Rank Island about five years ago, and it's all been reshaped. For the most part, unless I pointed it out, no one would know that the place had ever been mined. We're still irrigating out there to get the native plants started for the first couple of years. Ultimately they will be weaned and will be forced to operate off natural rainfall. You're going to lose some of them, we all understand that, but natives are supposed to flourish on their own.

People are very protective of their own property and their private property rights. When there was first talk about a Parkway going in along the river, I remember the landowners saying, "You can't come down here; this is private property. You want to take our rights away and let the nasty, unwashed public come down on our land." They were afraid there were going to be vandals and trespassers. Actually, most of the landowners have turned out to be very good stewards of the land. We wouldn't have a Parkway at all if it wasn't for them. Each landowner, in his or her own way, was an environmentalist, even if it's in a different sense than we usually think about.

I remember very distinctly the first meeting where I saw members of Save the River, or the River Committee—I can't remember exactly what they were called at the time, but they all showed up wearing buttons. We looked around asking, "What is this? Who are these kooks?" There were two things that were at the top of their list. One was to stop development on the river bottom. The second was to get the sand and gravel people out of there. At that point I knew that we had to make people aware of what really was happening on the river bottom. I know it hasn't always been a good association. In the past there have been mine sites—Lost Lake Park is one—that were never reclaimed. That material was used to construct part of Friant Dam in the late forties; it was a government project and no effort was put into reclamation.

There was some gravel mining before the dam was built. Actually, the San Joaquin River has always been the source of construction materials for Fresno and Madera. Mining the river bottom goes back a hundred-plus years.

I've worked on more than 2,000 acres of reclaimed sand and gravel sites along the San Joaquin River. That's not even counting the Ball Ranch, and the most recent acquisition was the Gibson property; that's another 325 acres. In exchange for allowing mining on their property, landowners get several things: royalties from mining—a significant amount of money—and a newly created habitat with lakes and ponds, all at no cost to them. The San Joaquin River Conservancy, a state agency, is getting all that now from the Gibson property, which they recently purchased. It's a long, narrow piece that has some beautiful river frontage. The conservancy is going to make a couple million dollars off of royalties that they can use for maintenance of the land.

The gravel companies do find gold, but not a whole lot. The bigger stuff drops out upstream as it comes down from the mountains; farther downstream, the gold nuggets keep getting smaller. What remains is basically gold dust that's mixed in with sand. The companies recover it, though, by running sand over indoor/outdoor carpet. The gold settles down into the carpet fibers; periodically they come in, peel the carpet out, roll it up and process it in a lab offsite. It helps pay the electric bill, but that's about it. Of course, that's nothing to sneeze at!

I've discovered that if you can convince a company that an action is not only environmentally correct, but also efficient and economically feasible, it will take that course. Initially it will do it because it makes economic sense, but gradually it will do it for the environmental purpose. Terms like "sustainability" were bad words just a few years ago; now they are very common practice.

When I first started working on the Underdown property, I began to notice the abundant wildlife along the river. I was photographing the progress of the reclamation project, and I started to notice that my pictures were inadvertently picking up wildlife. As time went on, I'd come back from the field and realize that I hadn't taken any pictures to document the reclamation. I'd walked through that whole site and not gotten one shot of the entire project—just the wildlife.

I've been fortunate enough to get a couple of pictures of the albino deer. The first one I saw was in the mid-nineties. There is always a small deer herd below the dam, but in the wintertime it gets larger when more deer come down from the foothills, have their young, and go back up. Somewhere in that genetic pool there's a powerful albino gene; there have been at least six different albino deer that I know of. It seems like about every two years there's a new one.

I made a mistake one time, though. Somebody called me and said, "I want to do an article about the albino deer." He mentioned the Department of Fish and Game, and so I told him the whole story and sent him one of my photographs. I told them I'd like to see the article when it was published. When I got the article, I realized that it wasn't a Fish and Game article at all. It was a hunting article. The article told a story about how the albino deer were special to Native Americans, that they were considered spiritual, which was pretty cool. But the end of the article said, "You should also know that the meat tastes the same whether it's an albino or a normal deer." Ah! The article also talked about how the albinos were really easy to spot in the fields, to get sight of. Believe me, I have never sent another picture out without seriously checking out how it was going to be used.

As you can imagine, some of the work I do is not much fun. I sit in an office at the computer; I have to deal with politicians and complaints from the public. People want cheap gravel but they want it to come from somewhere else—not from their backyard. My reward is to see the acquisitions and reclamations happen, to see the plants and wildlife coming back. My trips out into the field are my biggest payoff. I take long walks with my camera, see all the reclaimed habitat, and watch the evolution over the years. That's truly heartening.

"Yes, the river is majestic, the river is peaceful and life-giving, but it's also fun." —*Christina Ortez. Courtesy of Gene Rose*

Christina Ortez

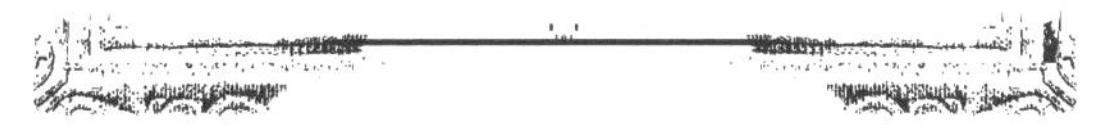

A Latina environmental activist, Christina Ortez is passionate about involving Latinos in protecting natural resources. She earned a degree in anthropology at Harvard and has had positions with several environmental protection organizations, including the Natural Resources Defense Council. She now lives in Albuquerque, New Mexico, and works with the Sierra Club as Southwest youth representative on the club's program Building Bridges to the Outdoors.

I GREW UP IN MADERA, way out in the country. I spent a lot of time on the river with my grandma, and we kids would go to Skaggs Bridge and swim. But I developed this huge fear of water from the time I was about ten years old, after a scary incident in Bass Lake. I felt really creepy about getting in the water, so that sort of stopped my connection to the river, and at some point I completely stopped going. Then I left the area for twelve years and stopped doing all the things that were connected with the outdoors. But I started feeling a deep need that I had never felt before, a strong pull to come back here to the valley, to come back and protect its resources. I didn't know where this was coming from at all, but the feeling was strong, so I came back and I met these great folks who helped me realize that it's really about our water, here in the Central Valley. It's all about water.

As a Latina, I could see that there weren't many Latinos and Latinas who were connected to the fight to protect water. I know that Latinos have such a strong connection to the San Joaquin River and to the Kings and Tuolumne Rivers, and that the feelings are very deep and very real. But emotions are not always connected to action.

When I was a kid, we went to Lost Lake all the time—we'd have barbecues there on Sundays. But it never really occurred to me that this was a resource that was important to who we were. You know, it's just something we *did.* And it's what a lot of Latinos still do every weekend. The people who grow up around here, especially the folks who don't have a swimming pool and don't have money to belong to a health club, we use these bodies of water for our recreation. We go to the beaches, we go camping and picnicking, but it's something that most of us don't really think about. Even though the Public Policy Institute shows that Latinos vote very strongly for park bonds, the action isn't there in other ways. The emotion that Latinos feel toward nature isn't something that's been formalized into activism. That's something that I'm trying to change. I want to make that connection between enjoying and recreating, between voting and being engaged civically.

So I started to bring together different folks to look at how we could be more involved in the process of protecting the environment and being good river stewards. I got a group of people together and we're talking about the ways we can bring Latinos into this scene. How can we integrate the fight for these resources? We want to add the voices that have been lost, add voices that haven't been listened to in the past or haven't been acknowledged because they were silent. With the increasing Latino population in California—it's 34 percent now—we are becoming stronger politically. But we need to educate ourselves about every part of California and its politics, and in this area so much of politics is about water. We're saying, "Start here! This river—from Lost Lake down to Skaggs Bridge—this is your river."

It took a lot to get me back on the water; it took someone saying, "You protect these rivers. You think about them a lot, intellectually. You think about how important it is to protect watersheds. But now you have to get back on the river." So I became a canoe guide, and then last year I decided that I would become a whitewater guide, too, which may seem ludicrous, given the real fear I had of being in the water. But now I can play on the river again. I'll tell you, it gives me a sense of fun like nothing I'd felt since I was eight years old, playing on the San Joaquin at Skaggs Bridge. I haven't felt that sense of freedom in a long time.

A couple of weeks ago I guided my first canoe trip down the San Joaquin. The peace I felt on the river was something that I had been looking for. The birds were out; the sun was setting. As we rounded the bend, we could hear a little bit of Highway 41, but that's okay because all I really noticed was the sound of birds. And we saw deer. That was shocking. What are they doing all the way down here? There were two kids in another canoe and they were so excited. We found a stick that had been eaten on the ends by beavers and that was a thrill, too. The feeling after a beautiful ride: it's heaven. That's something that everyone should experience at some point. Yes, the river is majestic, the river is peaceful and life-giving, but it's also fun. We need to get more and more Latinos on the river—for some fun.

I love to look at the valley's history—how the waves of migration have shaped our history and created what we are today. This valley is named after a saint because of Moraga, the Spanish explorer who came through the Central Valley and named many of California's rivers. He named this river, too. They think that it's because he crossed this river with his band of explorers on the feast day of Saint Joaquin, the father of the Virgin Mary. I don't think many people know that.

The story of the bandit Joaquin Murrieta is hugely popular within the Mexican community. We look to him for power and a certain *joie de vivre*. He gives us strength. Other myths play into our connection to the river, too. There is a story that is told to many Mexican American children, the story of La Llorona. She's "the crying lady," that's what *llorona* means. She was very depressed and she murdered her children in a body of water. They say it may have been the San Joaquin, or Cottonwood Creek, or it may have been Hensley Lake. Her spirit is always found near the water at night as she goes to look for the souls of her dead children. So if you're crying at night, your big brother or sister or cousin may say, "Don't cry, because La Llorona will hear you and be able to find you." It's a pretty scary story that is told to kids to make them stop crying at night. It's interesting, the myths that keep getting passed on and on. We stay connected to these places, like the river, through these kinds of stories, stories that seem to have something to do with us.

Our area is growing and growing. We've got these farms that need water but we've got this growing population, too.

How do we serve both needs? We need to understand the politics of water and rivers and dams. Friant Dam is necessary, but how do we protect the habitat below the dam? Balance is critical; I know it sounds like a cliché, but it's true. This valley would not be anything without farming. Many Latinos rely on farming for their livelihood. In fact, the whole valley does. There's got to be a way to make this all work through better technology, better water practices, by saving water and by not polluting our watersheds. There's got to be a way.

It is difficult to show why it's important that Latinos are part of the solution, why what we think matters, why we need to be involved in this fight to protect public lands and to keep the rivers flowing freely. It's a fight that's very challenging, but I like a good ruckus. I think it's a challenge I'm ready for, and I think the Latino community is ready too.

Lue N. Yang

A refugee from Laos, Lue Yang is now the executive director of the Fresno Center for New Americans, a nonprofit agency that serves more than ten thousand immigrants annually from all over the world, offering classes on citizenship, language, and employment. Bringing a completely different kind of fishing experience with him to the San Joaquin River, Lue tells about his own reeducation and transformation into a California fisherman.

I CAME TO THIS COUNTRY AS A REFUGEE from Laos. I first came to Honolulu and then moved to Des Moines, Iowa. I lived there for six years until my job transferred to Fresno in 1987, and I've lived here since then.

I had fished back in my old country, but we have very different techniques of fishing there. We sometimes used nets. We do not have the reel on a fishing pole; we just put string on a long bamboo stick. Use bamboo for a fishing pole and you can really throw far—many yards in front of you. There are all kinds of fish in Laos but they are completely different than in this environment. Only a few are the same as in the United States, such as suckers and crawfish. There is a type of fish in Laos that does not bite any bait; they only suck the fungus from the rocks. Fishermen drown them, they use a basket to pick them up or use their bare hands, and chase them into the hole and grab them. We have many different techniques in Laos.

When I moved to Fresno, I immediately learned about the San Joaquin River. My cousin took me to a picnic at Lost Lake. At that time I saw many people fishing. I came back to the river the next week, but I did not know what lures or bait could be used. I watched people using night crawlers, power bait, all different kinds, and that's how I learned to fish here.

One gentleman showed me that the Super Duper lure is very useful if you know how to use it. You put it on your fishing line, throw it out, bring it back, and the Super Duper twists back and forth and attracts the fish. At that time the river had all kinds of fish: trout, suckers, crawfish, and bass. You threw out your Super Duper and it was always easy to catch fish. But now many of those fish are gone.

I've fished on many different parts of the river. I've fished where the river comes into Millerton Lake. You can take a boat up there or go up to Auberry, where there's a road that goes down to the water. I think they call it Squaw Valley. Up there, if you're lucky, sometimes you can catch a big 45-inch striped bass. Very big.

Some people fish just for fun; they catch and they release. But a lot of people really like to eat fish, and when they catch a fish, they take it home and cook it. I like to eat fish. I always keep my limit

"Back in my old country, people normally used nets to throw over the fish....New immigrants have to learn how to use a reel and bait."—Lue Yang. Courtesy of Al Kawasaki

"The fish are not only for us; we have to save them for our children and our grandchildren." —*Lue Yang. Courtesy of Al Kawasaki*

and I will find a way to cook them the best I can. We cook similar to the Chinese style. We can do steamed, deep fried, baked, and also a style we call *la*. You remove all the bones, slice the fish into small pieces, squeeze lemon juice on the fish, and then separate the juice from the meat and add all kinds of spices. Stir it together and it tastes very good.

My wife is also from Laos and she knows how to fish too. She's a very good fisherman; sometimes she even catches the first fish of the day. There are a lot of people from my area that fish along the San Joaquin, not only Laotian but Hmong, Cambodians, and Vietnamese. We have a lot of people sharing the river. People move here and they already know how to find food in a river, so they immediately find out how to get access to the San Joaquin. New people coming here have to learn to obey the regulations to avoid receiving citations.

The first thing you have to do is get a fishing license, and learn what species and size and quantity you can take. If the fish is too big or too small, you have to let it go. Some places don't allow the use of live bait at all, you can only use a lure. Wax worms and grub can't be used in certain areas. Back in my old country, people normally used nets to throw over the fish, but over here it is illegal. New immigrants have to learn how to use a reel and bait. There are a lot of new things to know.

I sometimes take new people to show them how to fish and where to find the best place to catch something. I like to help and educate. When I first arrived in the U.S., I already spoke some English, so I had refugees coming to me for help. Later I became a caseworker for the refugee population and that's what I've been doing for thirty years. Now I'm the executive director for the Fresno Center for New Americans. I've overseen this organization since 1992. The funding comes from the Office of Refugee Resettlement in Washington, DC, to provide employment services for new arrivals. We have also been able to add mental health services, social services, health education, ESL classes, citizenship and voter education classes, and cultural education. We serve all refugees, including Southeast Asian—Hmong, Cambodian, Lao, Vietnamese—Ethiopian, Armenian, Russian, Bosnian, and Latin American. We serve some of the Burmese refugees that recently came from Thailand. We just began to see a few Iraqi refugees; that's the newest group.

I'm very busy, but I do have my weekends and holidays free, and that's when I spend my time fishing. I love to catch fish, yet I know we have to protect them. The fish are not only for us; we have to save them for our children and our grandchildren. If we take everything now, soon there will be nothing left.

Dora Canales

Dora Canales had been bringing her family to the San Joaquin River at Lost Lake Park for over thirty-five years, but it was not until her recent purchase of the well-known Dam Diner that she became an unofficial spokesperson, historian, and advocate for the river village of Friant. She told her story at the exact spot where she and her children used to swim in the river, and she brought with her faded snapshots—evidence of her lifelong love for the river.

When I bought the Dam Diner in Friant, I didn't know what in the world I was doing. I had never worked in a restaurant in my life. Never. My daughter Melissa tells everybody, "The day after my mom bought the diner, she went to the bookstore and bought *Restaurants for Dummies*." And I did. I just love that book. I refer to it often, even now.

I was a state-certified court Spanish-language interpreter for over thirty years. I'm still doing some of that—private hearings, cases, depositions. I did a lot of training because it's not just a matter of knowing Spanish. We're talking about people going to jail in murder and federal cases, so you have to be very precise. Now I'm more like a clearinghouse. The courts call me and I'll send them somebody to interpret. If I can't find anybody, I'll run over there at ten and cover it. Two hours later I'm back at the diner.

When my kids were little we always used to come up to Lost Lake. It's where we came for picnics and family things, and sometimes we'd go up to the dam. I have photographs of us by the dam from thirty years ago. We always came to this river; the area is so beautiful. My daughters loved to swim and play on the rocks; we used to come here whenever we could. It was good for us to be in the fresh air, in nature. The same rock that I used to sit on is still there. Nothing has changed. It's just as beautiful as ever. I have photographs of my daughters on those same rocks. The picnic table we used looks exactly like the ones there now.

The diner itself has been here since 1947. We used to go there all the time, but I hadn't been in there for many years. One day, out of the blue, I made plans to meet there with my son-in-law and my daughter Melissa. We met just to eat lunch, and a week later I bought the Dam Diner.

I don't know exactly how we found out the diner was for sale. My daughter saw it listed on the Internet. They were the ones who were supposed to run it because Ty, Melissa's husband, had worked in an Italian restaurant, and Melissa had managed a Fuzio and some other restaurants. So I was the one who purchased it, but they were going to run it. I was supposed to just come by on weekends and give my cooking tips. Then, all of a sudden, Ty got called to Iraq. He's an Apache and Blackhawk

"My daughters loved to swim and play on the rocks; we used to come here whenever we could. It was good for us to be in the fresh air, in nature." Courtesy of Dora Canales

Dora Canales, pregnant with Melissa, and Marianela pose with Friant Dam in the background, circa 1975. Courtesy of Dora Canales

pilot, so they shipped him out. He did three tours. The first two were short, but the third one was ten months. My daughter was a nervous wreck waiting for Ty to come back. He did make it back safely but he was gone for a long time.

Now they do come to the diner sometimes and help as much as they can. But it's my place, really. I finish a deposition and the next thing I know, I'm at the diner. Tomorrow, deposition at ten o'clock; as soon as I finish with that, I'll be at the diner.

For two years after I bought the diner, I stood at the entrance to the kitchen, watching. I looked at the menu, wondering, "How do they do all this?" To me it was shocking that the cooks could prepare different things at the same time, but they had their system. I stood there watching them and they would turn around and say, "What are you doing here? You should be out in the front." I ignored them until I learned everything. Their hamburgers are excellent; their omelets are excellent; but I was surprised that they only had a little page of Mexican food. Nobody said anything about it to me, but slowly I would take an item out and put another item in. I'd say, "No, you don't put enchilada sauce on the rellenos." I changed things slowly but surely, and we've perfected everything. When I first started, my daughter would say, "Chill out, Mom." I'd get very nervous, saying, "How long has that been in there?" and "What's the temperature?" I was being cautious but, of course, you have to be careful. Now I can relax because at least I have a little knowledge of the business.

I've met many interesting people at the diner and I love history. People have told me so many stories. Many of my customers have been living in the area for years. People say they've been coming to the diner their whole lives. There's a gentleman who comes in often. He says that he's been coming to the diner since he was nine years old, and he's sixty-plus now. He said they used to sell through the window and they would trade fish for a hot dog. At that time, the diner was named Steve and Jerry's. He brought me a picture from when they used to sell from the window.

Old-timers talk about how there used to be a store over by the canal that wouldn't let the Indians in. They talk about the stagecoach that used to pass through there. They showed me where the marks are, where it used to pass. The old hatchery used to be right there next door. You can still see it, but it's all boarded up. The railroad was right behind the diner. One gentleman who comes in told me that he has one of the railroad cars in his garage.

We get the fishermen as well as the people who hunt, mine for gold, play golf, and swim and canoe on the river. People come on their way to and from the casino. That's why we serve breakfast all day, because they'll stop in anytime. Sometimes they'll come in at four o'clock in the afternoon, having been up all night, and they want breakfast. I say, "Okay, let's pull those eggs out again."

Everybody knows about the Dam Diner. As a matter of fact, one night while my son-in-law was in Iraq, the bombs were going off all over the place, and he was trying to go on the Internet. He searched the Dam Diner and it came up. He turned to all his friends, "Hey, look at this!" He said all he wanted to do was come home and be a cook.

People know about us. A gentleman from the *Sierra Star* in Oakhurst wrote about the Dam Diner, but mainly about the fishing on the river. A gentleman from the *San Francisco Chronicle* wrote about the bald eagles in 2004 and said that he ate at the Dam Diner and loved it. But mostly he talked about the eagles. Other people come here because of the birdwatching, and other people come for the flowers. The motorcycle club meets every third Wednesday of the month; they have their meeting in the diner. There's a group that comes to the river to paint every Tuesday, and then they come to the diner after they finish their paintings. The Corvette owners have a Cobra Club. Those cars are gorgeous—fifty Corvettes in our parking lot. The Red Hat Society comes in all the time too; they are the ladies in the purple dresses and red hats. They've sent me letters saying that they love the atmosphere here.

Golfers have always gotten 20 percent off their bill if they bring in their scorecard. The custom was established many years ago. They get out of the car and say, "Get the scorecard!" I didn't know anything about it at first. I'd see them attach their cards to the tickets but I didn't know what I was supposed to do. They had to teach me. Not too long ago, some people that were coming back from skiing said, "Hey, we have our scorecard. Do you take the lift tickets from Sierra Summit?" I said, "No, but that's an idea I'll think about." The golfing discount has been around forever, but I'll have to consider the skiers. I have to think about what makes a better business. The fishermen want me to open earlier because fishermen go fishing at dawn. Can you imagine when salmon are back in the river? I wonder how much business will change if the salmon come back?

I don't know how or why the restaurant started collecting the James Dean memorabilia. Most of the memorabilia is about James Dean, but there are a few things of Marilyn and Elvis. People give me so many things of James Dean. We have the speeding ticket that James Dean got on the day he was killed. Another man brought me some pictures of James Dean with the car that he had crashed before, in the streets in L.A.

Last summer a very nice-looking gentleman came in. He had boots on and a nice cowboy hat. He was looking at the pictures of James Dean and reading the article about his last autograph. He said, "James Dean was my first cousin; my mother and James Dean's mother are sisters." He recalled riding in the car to the scene where Dean had died; his mother and aunt had cried all the way. The man said that his mother, James Dean's aunt, lives in the trailer homes behind us, and we don't even know who she is. She might come in and we'd never know. Another lady came in once and said, "I'm Sissy, Sissy King, from *The Lawrence Welk Show*. I'm the one who danced."

She lives in Arizona but she came up here for lunch with a friend.

I just got a framed poster of the cyclist Lance Armstrong. This was the first time Amgen Tour of California passed through here, so I framed it. A lot of cycling people stop at the Dam Diner, and they were surprised that we weren't notified that the race was going to be passing at a certain time. They were surprised that the posters listed all the towns that the race went through, but Friant was not listed. Somebody should have notified us. Once people started finding out, they were all out there watching; the whole diner came out and there were lots of people in front of the post office with their chairs set up. It was exciting.

When I first bought the diner, a couple asked to speak to the owner. I said, "What about, good or bad?" They said, "We wouldn't mind paying more if you had some better wine." You get all sorts of people out here, and when the food is good, you get repeat customers. Some businesspeople offered me a chance to split the lot up and install a gasoline station and a mini-mart. I said, "No, we don't need another gas station." There are mini-marts everywhere and there's no personality in that. So I'm hanging in there. If I sold the diner, maybe my own troubles would go away, but I can't do that. People would pass through town and feel terrible. It would be very sad.

This is downtown Friant! It's a landmark. We need a good restaurant in Friant. In fact, I've been going around to other restaurants and thinking, *What kind of restaurant do we want to be?* Maybe something like the Daily Grill in Fresno; that's a nice restaurant. They don't have biscuits and gravy, but they do have a diner-type setting and a patio like us. Maybe someday, here...Friant should be known for something great. Of course I know it already is—the eagles, the birds and flowers, and the great San Joaquin River.

Acknowledgments

Many thanks are owed to the individuals and institutions who made this book possible. First, we cannot imagine having finished this book without the amazing support of our assistant editor, Donna Mott. She volunteered hundreds of hours of hard work, creativity, phone calls, organizing, researching, begging, borrowing, and sleuthing. She was a marvel. Phoebe Farnam was our devoted copy editor, volunteering many hours to properly punctuating, structuring, and cutting these stories. Jacalyn White has been with us since the beginning, offering her video editing and technical skills to the interview stage of the project, and later lending her keen eye as one of our last readers. We also thank our very last reader, Susan Currie Sivek, for putting the final pencil to these narratives.

For sharing their life stories, we want to thank all those narrators whose interviews appear in this book, plus those we were unable to include: Charles Baley, John Baltierra, Phyllis Baltierra, Duayne Barker, Genevieve Barrangan, Tony Barrangan, Elmer A. Belmont, William Bertolani, Emiliano Chavez, Lois Conner, Congressman Jim Costa, Scott Davison, Bob Davis, Marveda DeBoer, Duane Furman, Donna Greeno, Adams Holland, Glenn Holly, Tom Jeffus, Marvin Kientz, Dave Koehler, Herb Lyttle, William Louie, Dr. Pete Mehas, Carlos Navarrette, Jim Oakes, Robert Pennell, Patricia Perry, Dave Phillips, Annabel Pitman, Ruby Pomona, Cathy Rehart, Bertina Richter, Gene Rose, Dr. Marcia Sablan, Henry Shein, and Grace Tex.

For their various forms of additional support throughout this process, we thank: the entire Parkway staff, especially Dave Koehler and Kathy Lustig; our most diligent and dedicated transcriber, James Wilson, and Natasha Ruck for additional transcriptions; Anna Wattenbarger, Bill Wattenbarger, Jane Campbell, and Anidele Flint for additional proofreading and fact-checking; Billy Seacrest from the history department at the Fresno County Library for his assistance in finding historical photographs; Peerless Pumps and the Eastern Fresno County Historical Society for further assistance in finding photos; and Rich Milhorn and Dave Grubbs for providing high-quality photo scanning. And for their various miscellaneous tasks and accidental functions, we also thank: Elise Moir, Roy Ramsing, and most especially, Tricia Garlock and James Hallowell.

Finally, many thanks to Malcolm Margolin of Heyday Books for his confidence in this project, and to Gayle Wattawa, our acquisitions editor at Heyday, who patiently led us through the publication process.

About the Editors

Joell Hallowell is a filmmaker, writer, and photographer living in San Francisco. She was recently an assistant editor and interviewer for *Underground America: Narratives of Undocumented Lives*, an oral history project of the McSweeney's Voice of Witness series. Her collaborative films have been screened at various experimental festivals and venues, including the Harvard and New York Film Archives, the London International Film Festival, the Madcat Women's Film Festival, and the Chicago Underground Film Festival. She is currently editing two projects, *Here Come the Brides*, a book of oral histories and photographs that capture the stories of twenty couples who were married in California during the few months in which gay marriage was legal, and *Shadowed: Disappearing Women*, an anthology of contemporary poetry written for, and paired with, old photographs of anonymous women from the first half of the twentieth century.

Coke Hallowell grew up in the San Joaquin Valley and has been involved in conservation issues for twenty-five years. In 1986 she was part of a grassroots group of citizens, the San Joaquin River Committee, who were concerned about the fate of the San Joaquin River and rallied to protect it. In 1988 the San Joaquin River Parkway and Conservation Trust was formed with Coke as president, a position she retained until 2008, when she was elected to her current position of chairman of the board. She was a founding member of the Sierra Foothill Conservancy and the California Land Trust Alliance and has served on the Environmental Equity Committee and the Resources Legacy Foundation Board of Directors. She currently serves on the Planning and Conservation League Foundation Board of Directors.

About Heyday

Heyday is an independent, nonprofit publisher and unique cultural institution. We promote widespread awareness and celebration of California's many cultures, landscapes, and boundary-breaking ideas. Through our well-crafted books, public events, and innovative outreach programs we are building a vibrant community of readers, writers, and thinkers.

Thank You

It takes the collective effort of many to create a thriving literary culture. We are thankful to all the thoughtful people we have the privilege to engage with. Cheers to our writers, artists, editors, storytellers, designers, printers, bookstores, critics, cultural organizations, readers, and book lovers everywhere!

We are especially grateful for the generous funding we've received for our publications and programs during the past year from foundations and hundreds of individual donors. Major supporters include:

Anonymous; Audubon California; Barona Band of Mission Indians; B.C.W. Trust III; S. D. Bechtel, Jr. Foundation; Barbara and Fred Berensmeier; Berkeley Civic Arts Program and Civic Arts Commission; Joan Berman; Lewis and Sheana Butler; Butler Koshland Fund; California State Coastal Conservancy; California State Library; California Wildlife Foundation; Joanne Campbell; Keith Campbell Foundation; Candelaria Fund; John and Nancy Cassidy Family Foundation, through Silicon Valley Community Foundation; Christensen Fund; Creative Work Fund; The Community Action Fund; Community Futures Collective; Compton Foundation, Inc.; Lawrence Crooks; Ida Rae Egli; Donald and Janice Elliott, in honor of David Elliott, through Silicon Valley Community Foundation; Evergreen Foundation; Federated Indians of Graton Rancheria; Mark and Tracy Ferron; Furthur Foundation; George Gamble; Wallace Alexander Gerbode Foundation; Richard & Rhoda Goldman Fund; Evelyn & Walter Haas, Jr. Fund; Walter & Elise Haas Fund; James and Coke Hallowell; Sandra and Chuck Hobson; James Irvine Foundation; JiJi Foundation; Marty and Pamela Krasney; Robert and Karen Kustel, in honor of Bruce Kelley; Guy Lampard and Suzanne Badenhoop; LEF Foundation; Michael McCone; Moore Family Foundation; National Endowment for the Arts; National Park Service; Organize Training Center; David and Lucile Packard Foundation; Patagonia; Pease Family Fund, in honor of Bruce Kelley; Resources Legacy Fund; Alan Rosenus; Rosie the Riveter/WWII Home Front NHP; San Francisco Foundation; San Manuel Band of Mission Indians; Deborah Sanchez; Savory Thymes; Hans Schoepflin; Contee and Maggie Seely; James B. Swinerton; Swinerton Family Fund; Taproot Foundation; Thendara Foundation; Lisa Van Cleef and Mark Gunson; Marion Weber; John Wiley & Sons; Peter Booth Wiley; and Yocha Dehe Wintun Nation.

Getting Involved

To learn more about our publications, events, membership club, and other ways you can participate, please visit: www.heydaybooks.com.